the book of gods & goddesses

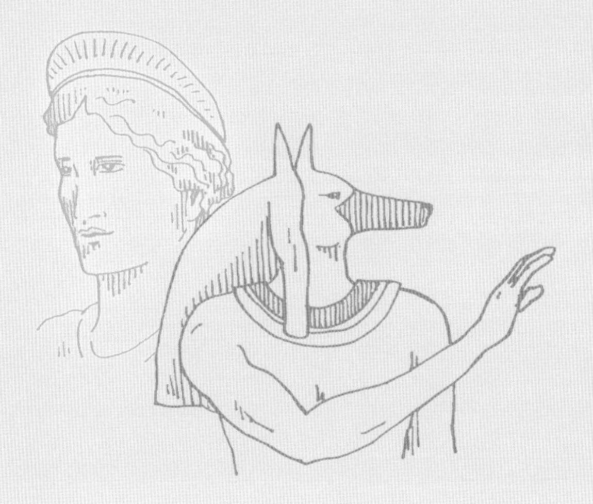

the book of gods & goddesses

a visual directory of ancient & modern deities

HarperEntertainment
An Imprint of HarperCollins*Publishers*

HarperCollins books may be purchased for educational, business, or sales promotional use. For information please write: Special Markets Department, HarperCollins Publishers Inc., 10 East 53rd Street, New York, NY 10022.

FIRST EDITION

Library of Congress Cataloging-in-Publication Data has been applied for.

ISBN 0-06-073256-3

04 05 06 07 08 10 9 8 7 6 5 4 3 2 1

Conceived, designed and produced by Quid Publishing
Fourth Floor, Sheridan House
112 Western Road
Hove BN3 1DD
England

www.quidpublishing.com

Publisher: Nigel Browning
Design: Lindsey Johns
Illustrations: Matt Pagett
Picture Editor: Veneta Bullen

Printed and bound in China by Midas Printing
All photographic images courtesy of Corbis

contents

introduction

Right: Venus – the Roman goddess of love, known as Aphrodite in Greek mythology,and inspiration to many great works of art.

Below: These beautiful statues (caryatids) support the roof of the Erechtheion Temple in the Acropolis (420 BCE) and overlook the modern city of Athens.

"As flies to wanton boys are we to the gods" are the words Shakespeare gives to Gloucester in *King Lear*. But in reality, it is we humans who make and unmake the gods. Many of the divinities in this book are no longer worshiped. Their temples lie in ruins, the haunt of tourists and scholars. But the mighty gods of old have not perished. They live on in our daily lives and in our everyday language. Thursday is the day of Thor, the Norse thunder god; we visit museums, named for the Greek spirits of culture; and we use words such as *venereal* (from the goddess Venus), *erotic* (from the god Eros) and *Hell* (from the Norse goddess Hel).

Left: Relief sculpture (dating from the fifth century BCE) depicting Poseidon, Athena, Apollo, and Artemis.

Below: In Norse mythology, Thor was the god of war, thunder and strength. The son of Odin, he destroyed the enemies of the gods with his magic hammer.

Below left: Hades is the Greek god of the underworld (Pluto in Roman mythology). Black sheep were regularly sacrificed in his honor.

Many of the gods and goddesses found in this book have passed into history but they have not died. Quite the reverse, in recent years they have made something of a comeback as the followers of New Age movements use their mythologies as a means to discover an inner spirituality: the god or goddess within.

This book is divided into seven sections, each exploring a separate region and its rich vein of myth. Many of these beliefs, such as the worship of the Olympian gods of Greece and Rome and the Norse gods, have vanished, while others, such as Hinduism and Buddhism, are living faiths followed by millions of adherents all over the globe.

Absent from the pages of this book are the world's three great monotheistic faiths: Judaism, Christianity, and Islam. Instead, I have chosen to concentrate on the much less familiar pantheons of the polytheistic faiths. In polytheism gods and goddesses have to share the world with one another. This makes the myths about them extremely rich and varied. The gods often behave like the cast of some divine soap opera: falling in love, marrying, having illicit affairs and quarrelling, often with tragic consequences, sometimes for themselves but more usually for the poor mortals caught up in their schemes. The gods can appear to be very human, which has led scholars to argue that they

Right: Great Buddha or Daibutsu in Kamakura, Japan. The bronze sculpture was created in 1252 and stands in the grounds of the Kotokuin temple.

should be viewed as no more than projections of the human psyche.

Although subject to our frailties, the gods can also do things that we cannot. They have powers and qualities we lack – whether benign or evil. They are superhuman and, almost without exception, immortal.

A few, like Osiris in Egyptian mythology, die and are resurrected; others, like the Greek hero Heracles and some of the Roman emperors, live as mortals and are deified after death. The Norse divinities have to eat apples to stay young, but eat apples they do. The gods do not have to suffer our own inevitable fate; they do not live with one eye upon the end of their lives. Death separates mortals from the gods.

The gods may be eternal, but they do have histories. As well as worshiping them, humans have told stories about them to explain their characters and the nature of our world. We have also appealed to them for help with our affairs. We have made offerings and sacrifices and done our best to keep their goodwill. In these pages the reader will encounter many different divinities, from fertility goddesses to gods of destruction. Some are now obscured by time, others still worshiped, and all remain powerful and enduring symbols of man's religious imagination, a testament to our need to bridge the gulf that divides earth from the heavens.

Left: Osiris, the Egyptian judge of the dead.

Below: Ganesh, the Hindu god of good luck.

Below left: Amaterasu, the Japanese sun goddess.

Below right: Chalchihuitlicue, the Aztec goddess of storms.

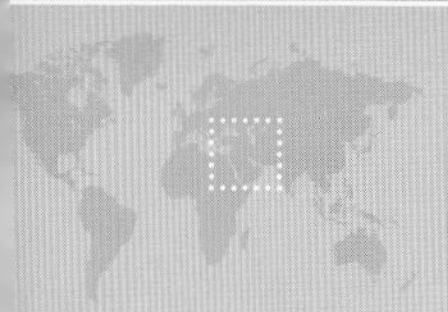

near east

Right: Nekhbet the Egyptian vulture goddess, guardian of mothers and children. With the rise of the pharaohs, she became the great goddess of Upper Egypt.

Below: The Grand Staircase at Apadana in Persepolis, one of the ancient capitals of Persia, near Shiraz in southern Iran.

The ancient Near East, encompassing the lands of Egypt, the Levant, Iraq and Iran, was the cradle of many of the world's great religions. Among the earliest civilizations to blossom there were the two riverine kingdoms of Egypt and Mesopotamia. The Egyptians relied upon the yearly flood of the Nile, while the Mesopotamians had to contend with the unpredictable Tigris and Euphrates. Forming a buffer state between Egypt and Mesopotamia were the kingdoms of Canaan and the rich trading ports of the Phoenicians. Farther to the east, on the great Iranian plains, arose a succession of mighty empires that ruled Asia from the borders of India to the Hellespont in Turkey.

The Mesopotamian gods changed considerably over time, often following the fortunes of the region itself. The Sumerians had settled in Mesopotamia by 4000 BCE and developed city-states. A thousand years later, they were writing on clay tablets, and their first myths were properly established by around 2500 BCE. Following conquest by the invading Akkadians, and then, around 2000 BCE, by the Amorites, who made Babylon their capital, the gods evolved from being nature powers to a more patriarchal pantheon in which each god had specific functions.

The Egyptians began to settle in the Nile Valley in around 3000 BCE. As rival cities grew, each favored its own god. As the fortunes of the cities changed, so did the functions and mythologies of their gods. Shifts in political power, especially between Memphis and Thebes, and extended periods of internal strife in the second millennium BCE, led to considerable changes in beliefs about the gods. Unlike the Mesopotamians, the Egyptians distinguished between good and evil gods. They also developed a notion of an immortal soul that survived in an afterlife. Not only did the dead become involved in divine struggles between good and evil but they would also be judged according to their behavior.

The lands of Canaan, which lay roughly where Israel, Lebanon, and Jordan are today, were a useful defensive frontier for the Egyptians. Between 1550 and 1070 BCE the Canaanites were under Egyptian control. *Canaanites* is a broad term, as it encompasses the coastal Phoenicians farther north in modern Lebanon. They were generally farmers who occupied the land crossed by the important trade route between Egypt and Mesopotamia, which also served as a highway for armies and shepherds. The Canaanite gods reflect the needs of a farming culture: fertility, rain and growth were all important concerns.

The lands to the east of Mesopotamia, covering modern-day Iran and Afghanistan, were the seat of the mighty Persian empire, whose prophet, Zarathustra (c. 1200 BCE) taught the worship of the God of Light, Ahura Mazda.

Top: The remains of a ziggurat (temple) at Ur in Iraq, the ancient capital of Sumeria and the birthplace of the prophet Abraham.

Bottom: The Pyramid of Khafre one of three located on the Giza plateau on the outskirts of Cairo.

GODS OF MESOPOTAMIA, PERSIA AND CANAAN

Ancient Mesopotamia is often called "the cradle of civilization." Sumerian culture emerged early in the third millennium BCE in the valley between the Tigris and Euphrate rivers, in what is now Iraq. The Sumerians invented cuneiform writing by inscribing tablets of clay around 3000 BCE. Mesopotamian culture and beliefs were extremely influential on their immediate neighbors to the West, the peoples of Canaan, and those t the East, the inhabitants of Persia, who would one day become their rulers.

Above: Much of Mesopotamia can be found today in Iraq and most of Persia is in Iran. The ancient land of Canaan is where Israel exists today.

The most important collection of traditions about the gods is *The Epic of Gilgamesh*, the earliest versions of which date from around 2600 BCE. Another valuable source of information about the Mesopotamian gods is the *Enuma Elish* (*When on High*), a Babylonian creation and flood narrative that may date from as early as 1700 BCE and which is closely related to the Biblical creation story in Genesis.

The ancient Sumerians believed that the gods controlled human destiny. In each of their cities, a temple was dedicated to one of their gods, to ensure that the relevant deity would look benevolently on them and intercede on behalf of the townspeople or even an individual family during times of need. Yet the Sumerians believed that there were no guarantees the gods would help them. Certainly, they did not expect their gods to reward pious behavior on earth with eternal joys in the next life. For the Sumerians and the later Babylonians the "House of Dust" awaited the soul of the dead a grim place of eternity where the soul, dressed in feathers like a bird, would feed on soil and clay, see no light and dwell in darkness.

Babylon emerged as an important city around 2000 BCE and the Babylonians adapted the Sumerian gods, making them patriarchal instead of matriarchal, and giving them more specific functions. The Babylon enjoyed close trade ties and a common ancestry with Canaan, which became the promised land of the ancient Israelites. Their mythology and religion also had strong links. Both cultures told very similar stories and shared a belief in an underworld that awaited the soul of the dead.

However, less is known about the beliefs of the Canaanites, who occupied what is now Israel and Syria and were called Phoenicians by the ancient Greeks. They wrote on papyrus rather than clay tablets and the paper rotted over time so information about Canaanite gods remained scanty until an archaeological find at Ugarit in Syria in the 1930s. The so-called Ugaritic texts, dating from around 1400 BCE, provide accounts of creation and important information about the Canaanite gods. The Old Testament is another important source: Biblical prophets such as Elijah and Hosea railed against the worship of Baal. Over time, the god of the Jews, Yahweh, usurped the roles of the Canaanite gods to become the only god of the land.

One of the earliest monotheistic faiths of the world, Zoroastrianism, emerged from the polytheistic beliefs of ancient Iran around 1200 BCE. This faith endures to this day in the beliefs of the Parsees of India, who fled persecution in their native Iran.

Right: One of the most violent and bloodthirsty of the ancient deities, the warrior goddess Anat was the sister, and possibly the lover, of the great god Baal. She enjoyed the ecstasies of both sex and war.

Enki God of fresh water

Enki sided with mortals to help them escape the flood brought about by Enlil. He was the god of fresh water (including lakes, rivers, marshes, springs, and wells) and lived in the underground waters that the Babylonians believed supported the earth. He emerged to instruct men in agriculture and irrigation and devised the order inherent in civilization. Enki had esoteric knowledge about the nature of life and could raise mortals from the dead. The Babylonians drew him as a seated figure with a long beard, surrounded by water. They also pictured him with a body that tapered into the form of a fish.

In one creation account, the inventor figure Enki designed the molds from which people were made.

Enlil God of the air

The Babylonians believed that wind came from the mountains, so they gave Enlil, the god of winds and storm, a mountain home. He was so magnificent that the other gods could not look at him, and his names include "Great Mountain," "Raging Storm," and "Wild Bull," as well as "King of the World," "Lord of the Land," and "Father." He brought rain, invented the agricultural hoe (and therefore farming), and helped to create mankind by working the soil with his hoe: the first people grew out of the earth he had broken. Yet Enlil, ruler of gods and men, the supreme god of the Babylonians around 2000 BCE, was also a destroyer. *The Epic of Gilgamesh* recounts how, angry that mortals were keeping the gods awake at night with their noise, he sent a flood to destroy humankind. Later, in the first millennium BCE, the rise of Marduk as supreme being led to Enlil being marginalized and he became identified as a figure of evil.

The word of Enlil is a breath of wind,
the eye sees it not.
His word is a deluge which advances
and has no rival.
His word above the slumbering skies
makes the earth to slumber.
His word when it comes in humility
destroys the country.
His word when it comes in majesty
overwhelms houses and brings
weeping to the land.
At his word the heavens on high are
stilled.

Babylonian Creation Epic

The impressive ruins of Enlil's temple at Nippur, around 150 miles south of modern Baghdad, were once the center of his cult.

Gilgamesh Mortal hero and underworld god

Sumerian and Akkadian scripts from Mesopotamia were decoded in the early nineteenth century, leading to the translation of the Gilgamesh epic.

The hero of the world's earliest written story and one of its greatest epics, Gilgamesh, king of Uruk, whose mother was divine, travels through the underworld and onward to the ends of the earth to meet Utnapishtim, the immortal survivor of the Great Flood, a counterpart of the Biblical Noah. Gilgamesh seeks to escape death and, although he fails in his challenge to defeat Sleep, the brother of Death, Utnapishtim gives him "the plant of rejuvenation" but it is stolen by a snake and Gilgamesh returns to Uruk, lamenting the fact that he ever embarked upon his journey. Yet he has become wise: *The Epic of Gilgamesh* ends with Gilgamesh admiring the great city walls of Uruk. He has confronted his own mortality and realized that the way to transcend death is to leave a monument that will outlast one's lifetime.

The twist in the tale lies in a genealogy of gods recorded on clay tablets, which makes it clear that after his death Gilgamesh became Lord of the Underworld. Gilgamesh strove to defeat his own mortality, not knowing that because his mother was divine, he would escape the inevitable fate of ordinary mortals.

Ishtar Goddess of love

Ishtar's cult enjoyed prostitutes at her temples and revered her for the emphasis she placed on pleasure and joy.

Ishtar, the sister of Shamash, was a goddess with two roles: the goddess of love was also the mistress of battle. This duality may have been due to a merger of two divine figures into a widely worshiped goddess who went by different local names including Inanna and Istar. In Phoenicia, she was known as the great goddess Astarte; the Greeks incorporated Astarte's qualities into the identity of their goddess Aphrodite. Descriptions of Ishtar in ancient Babylonian texts occasionally refer to her as being bearded, suggesting her warlike qualities or linking her with a male figure. As goddess of love, the Babylonians identified her with the evening star, as a war goddess she was the morning star. She was believed to be beautiful and voluptuous, irritable and intolerant. In Babylonian art, she was depicted as a naked figure wearing a cap with horns.

Marduk Bull calf of the sun

Originally a relatively obscure god of storms who also presided over agriculture during the second and third millennia BCE, Marduk became god of the city of Babylon around 1400 BCE, and supreme Babylonian deity, a figure of wisdom and victory.

The creation epic *Enuma Elish* tells of his rise to power in a civil war with the gods. He rode alone into battle on his storm chariot, presenting such an imposing figure that the opposition ran away, save for their goddess leader Tiamat, whom he killed, splitting her corpse in two to form the earth and the sky. Marduk then ordered the defeated gods to build a city in his honour and created humankind to help them: the mighty city of Babylon was born.

Marduk rose in power alongside the rising influence of the city of Babylon.

Shamash All-knowing judge

Shamash, the son of Sin, represented the light – and life-giving qualities of the sun. A god of justice, he defended victims and punished wrongdoers, spreading light wherever he went. The Egyptians associated him with their sun god, Ra, and his journey across the heavens was similar to Ra's: he crossed the sky in a chariot, rising from the east with a saw in his hand, to cut through any mountains that might obstruct his passage. At sunset, he returned to the heavens through the western door.

Shamash was the patron of travelers and in *The Epic of Gilgamesh* supported Gilgamesh in his heroic journey.

Sin Lord of the calendar

Enlil's son Sin was the Babylonian moon god. He was revered both as a fearless creator and the divine determiner of human destiny who judged events in heaven and on earth. The Babylonians wrote his name as "30," since the lunar calendar had thirty days; they also symbolized him as a crescent moon lying on its back. He was sometimes depicted as a virile bullock with impressive horns, indicating his strength and divinity. Lunar eclipses were regarded as portents of doom, a sign that Sin was battling with demons in the heavens. Although generally regarded as a benevolent figure, Sin had a shady past, having made the goddess Ninlil pregnant by raping her.

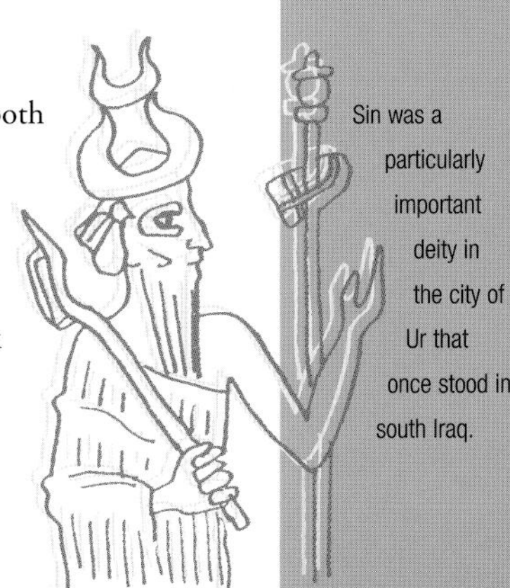

Sin was a particularly important deity in the city of Ur that once stood in south Iraq.

Ahura Mazda The wise lord

Once one of many gods of ancient Persia (Iran), Ahura Mazda was elevated to the position of supreme being and creator by the prophet Zarathustra.

According to the prophet Zarathustra (c. 1200 BCE), the founder of Zoroastrianism, the religion of ancient Persia, Ahura Mazda (Ormudz) was the supreme deity, creator of the gods, the divine powers, the world and humankind. Described as the sky-clad God of Light, Ahura Mazda is often depicted in the guise of the sun. The purifying element fire is his principal symbol and an important element of his worship in fire temples. During the Achaemenid period (558-330 BCE), he was represented as a pair of outspread eagle's wings. Although Ahura Mazda is all-powerful and will prevail at the end of time, he is opposed in all his works by the incarnation of darkness and evil, Angra Mainyu (Ahriman). One of the first faiths to abandon polytheism for monotheism, Zoroastrianism was an important influence on Judaism, Christianity, and Islam.

Baal Hadad Rider on the clouds

In the Old Testament (2 Kings, 2:2), Baal appears under one of his titles, "Baal-zebub" (Baal, prince of the earth). This title later became Beelzebub, a popular name for one of Satan's chief lieutenants.

Baal meant "lord" and the full name of the god referred to so contemptuously in the Old Testament was Baal Hadad; he was a rain god and fertility figure, the son of El. The Canaanites called him "Thunderer," "Prince," "Rider on the Clouds," and "the Strong One" and his cult was widespread.

Brother to the underworld god Mot and the sea god Yam, Baal fought them both to achieve sole kingship. He defeated Yam but desisted from killing him after their mother, El's wife Athirat (Asherah), intervened. Yam became a hostile figure, constantly threatening to flood Baal's kingdom.

Baal had a temple built for him, decorated with gold and silver. Mot entered it through a window, swallowed Baal whole and took him to the underworld. Baal's consort Anat rescued him and killed Mot. Until Baal was resurrected there was drought in the land but he emerged from the underworld to preside over the harvest. In ancient texts he is depicted holding a thunderbolt in one hand and a club in the other, with a crown on his head.

El Most High

In Canaanite mythology, El existed before all the other gods. A creator figure who fathered the other gods, he was the supreme deity. Following his act of creation, he became a more remote figure, an omnipotent god living with the other gods on his mountain heaven. When he heard that his son Baal had died and journeyed to the underworld, he came down from his throne (under which flowed the waters of the major rivers of the region), but, despite putting on sackcloth and cutting his cheeks in grief, he failed to rescue his son. El was always portrayed as a wise, bearded old man and his title "Father of Mankind" suggests that he was a god of the passage of time. He was also known as "Bull God."

El also simply meant "god," so every god was an "El." The Israelites used "El" as another word for "Yahweh," the great god of the Old Testament.

Mot God of death

The brother and rival of Baal, Mot was the god of sterility, death and drought. A giant serpent, he had a huge appetite for the living. He swallowed people whole and took them to the underworld he ruled. In ancient texts, he was poetically described as having a mouth so big that when open it stretched from earth to the stars. He entered Baal's temple and consumed Baal and his entire entourage, swallowing Baal like an olive. This great cosmic drama is a myth relating the cycle of the seasons: Baal, the fertility god, disappears, leaving behind him drought and desolation until his consort Anat slaughters Mot in an analogy of the harvesting of corn. She cleaves him with a sickle, beats him with a flail, burns him with fire, grinds him in a mill, then scatters Mot's remains over the fields, an act that ensures the fertility of the earth.

In the ancient texts, Mot is described as being so huge and powerful that when he consumes Baal, he does so with great ease, as though swallowing an olive.

Yam The sea god

Yam, god of the seas and rivers, was brother to Mot and Baal; the story of his battle with Baal is recorded on tablets dating from the fourteenth century BCE. Yam and Baal were both living with El. In a bid for power, Yam demanded that the gods deliver Baal to him, but Baal quickly defeated Yam and took him prisoner. He spared his brother when their mother, Asherah, reminded him that killing one's prisoners is a disreputable act, but from then on Yam regularly threatened to ruin the world with flood. He is the personification of the wild dangers of sea and river.

Canaanite temples contained a large basin filled with water to represent Yam.

EGYPTIAN GODS

The Egyptians worshiped many gods but they lacked a universal system of religious belief or a single comprehensive mythology. They had a number of myths that explained the world's creation, the nature of the gods, the origin of good and evil and the afterlife, and many of these contradicted one another.

Above: Maps of Egypt have not changed that much in history. The county is dominated by the Nile which flows North to its delta on the Mediterranean coast.

The Egyptians first settled as farmers in the Nile Valley around 3000 BCE. Before that, they had lived as tribes of hunter-gatherers. Each tribe had its own god, usually imagined as a spirit inhabiting the body of a bird or an animal; the Egyptians believed the world was structured in pairs of opposites: good and evil, light and dark, sun and moon, and order and chaos. They gave their gods identities that reflected their belief in these opposites; they had good gods and evil gods, gods of the sun and of the moon, gods of chaos and order. Increasingly, the gods were imagined in human form, though in drawings and hieroglyphic script (the Egyptian form of picture writing) they were often shown with the head of the animal they had first been associated with.

Like the Sumerians and Babylonians, the Egyptians believed in an underworld, a grim place where the dead spent the afterlife. Yet, especially from around 1500 BCE, they also developed the idea of a heavenly paradise. They believed that the gods involved themselves in the journey of the dead. Some of the gods helped judge the dead, some had the power to allow the soul to pass into heaven and others were evil influences. The deceased were equipped with spells and charms to protect their souls from evil and persuade the gods to allow them to travel on their journey. Information about the character of individual gods comes partly from these spells and charms, and also from the different cycles of myths.

Different Egyptian cities worshiped different gods and cult members would sometimes incorporate a myth about a different divine figure into a story about their own god. One of the main difficulties when encountering the Egyptian deities is that they often share the same roles and mythology and are therefore very similar to one another. This is more of a problem for the modern reader than it was for the ancient Egyptians, who were free to ignore one god and focus on another. Although there were many sun gods, it is unlikely that individual Egyptians would have worshiped more than one.

Right: The falcon-headed sun god Horus was widely worshiped in Ancient Egypt. Mythically, he avenged the murder of his father Osiris and was the ruler of Egypt.

Egyptian civilization faded around 400 CE, by which time the country had become Christian. In the nine hundred years prior to its decline, Egypt had been ruled by Persians, Greeks and then Romans, and had considerably influenced the world-views of its occupiers. The deities became a cultural export but until the translation of hieroglyphs in the early nineteenth century the modern world knew little about Egyptian gods and goddesses.

Amun The elusive god

Amun or Amon meant "the secret one" for mortals could not know his real name. His statue was shrouded with cloth to indicate his hidden nature. As one of the earliest Egyptian gods, dating from the third millennium BCE, Amun was originally god of the air and the wind (both invisible and elusive). Later, around 1400 BCE, following a militaristic era, Amun was adopted as the god of war and then as "king of the gods," the supreme creator who even created himself.

He appeared in human guise to impregnate the wives of the kings, ensuring the divinity of the succession. He also valued justice, healed the sick and protected commoners.

Always depicted in human form, generally bearded and with light blue flesh, Amun wore a crown of two plumes. The Egyptians often represent him with the disc of the sun on his head. He was closely associated with the ram, a symbol of fertility.

The fossil ammonite is so called because it looks like the horns of a ram, which was associated with Amun.

Anubis Reader of the scales

Anubis was one of the earliest Egyptian gods, the god of death, guide and protector of the dead. He had a human body but the head of a jackal, with a long muzzle and alert ears, and was often represented crouching down with his long tail curling out from behind and a magical collar around his neck. Before the ascendancy of Osiris, the Egyptians believed that Anubis judged the dead and ensured they would have enough food to eat in the afterlife. In the mythical tradition that assigned this role to Osiris, Anubis crouched next to the scales of justice, on which the heart of the dead was weighed against the feather of truth. He read the scales and gave the verdict to Osiris. Anubis was the patron of embalmers and the priests who presided over embalming wore jackal masks. According to a myth recorded by the Greek historian Plutarch, Anubis was the son of Nephthys and Osiris, but was raised by Isis.

The word "Anubis" probably came from the Egyptian word "to putrefy," which further linked the god to the process of decay, embalming and mummification that accompanied burial.

Apophis The evil god

Relatives routinely armed the deceased with spells to help them escape Apophis.

A real baddie: the serpent-dragon Apophis represented all the terrible forces of chaos and eternally aimed his lethal powers at the forces of good. Apophis had a terrifying roar that echoed throughout the underworld, and fed himself by breathing his own shouts. He lay in wait, his massive body coiled, to swallow anyone he could, mortal or god. Those he consumed became void, nothing.

Apophis could be subdued, but was very hard to defeat, almost always returning as a dark and hostile force, although in a version of one myth, the god Re destroyed him by hacking up his serpent's body and burning the pieces. More usually, he was portrayed as being chained by the neck to a gate of the underworld, restrained by the immortal forces of good and by magic, so that Re could pass by.

Atum The complete one

The mythical bird the Greeks called a phoenix was an Egyptian heron, an emblem of Atum. In some Egyptian myths, it created the cycle of time, which included the 24-hour day.

Atum's name came from the word "not to be" and meant "the complete one" or "the one not yet come into being." The Egyptians also called this creator god "lord of Heliopolis," which was an important center of sun worship.

Atum created himself, took the form of a serpent, and made the first creatures and then the whole of creation out of himself from the primeval waters of Nun. Later he took on human form. Other creation myths hold that the first creatures arose from Atum's seed when he masturbated, or from his spit. Atum was eventually absorbed into the figure of Re to become Re-Atum, a sun god.

The Egyptians linked Atum both to the beginning of the world and its end, when they believed he would destroy his creation and submerge everything, apart from Osiris and himself, in the primeval waters.

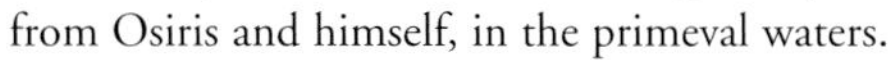

In ancient texts Atum is shown as a bearded man, wearing a twin crown and holding a staff of office. He was sometimes shown as an old man carrying a walking stick, representing sunset. He used to appear to Egyptians as a mongoose (an animal which preys on snakes).

Often believed to be the father of the kings of Egypt, and the source of their power, Atum helped raise the kings from the dead. As a manifestation of Re he battled and defeated Apophis in the underworld; he also protected the dead as they journeyed to the blissful paradise of the "field of dreams."

Bes The God of Music

Bes emerged late in Egypt's history and became one of the most popular Egyptian gods. A bearded bandy-legged dwarf, with shaggy hair, and a tail, he was often shown sticking out his tongue and wearing a lion's skin. His ugliness frightened off evil spirits and he used to bite the head of serpents, but was much kinder and less scary to mortals than he looked. Bes protected women in childbirth, and was the patron of music and merry-making, all of which contributed to his popularity. He also brought good fortune and wealth, and happily played both harp and tambourine. The Egyptians made numerous pendants and amulets in his honor. Because they linked him with all the joys of life, married couples often placed a statue or image of him at the foot of their bed, to ensure their happiness. Outside of Egypt, Bes was popular in Phoenicia, where he often appeared on furniture decorations. Artifacts alluding to Bes, dating from the second millennium BCE, have also been recovered in Cyprus. The Romans depicted him in a legionary's outfit.

Bes's popularity extended to the royal houses of Egypt and his picture was often painted on bedroom walls, perhaps to encourage sexual fulfilment.

Geb God of the earth

It is difficult to give the earth god Geb a good press, because he violently raped his mother Tefnut after falling in love with her, when his father (Shu) was absent. Geb was never punished for the act. In another ancient text, Geb acted as a prisoner of the dead, binding them within his body. Yet Geb had many positive attributes and was kind to mortals. He was often painted green, a natural color, to show that vegetation sprang from him, and was sometimes depicted lying under Nut (the personification of the vault of heaven) while making himself comfortable on a pillow, or as a goose-headed man. Egyptians therefore referred to him as "The Great Cackler."

Geb formed the earth, which to the Egyptians was flat, with his back by lying in the center of a circular ocean that stretched from the underworld to the sky. Geb was Atum's grandson and consort-brother of Nut, who gave birth to their son Osiris, whom Geb preferred to Seth as heir to Egypt's throne. Presumably because of his role in establishing the Osiris lineage, the kings of Egypt were known as the heirs of Geb.

When Geb laughed, the earth shook: an ancient explanation for earthquakes.

Hathor The celestial cow

Hathor once cured her father of a fit of anger by dancing naked in front of him until he laughed.

The daughter of the sun god Re, Hathor's many titles included "Lady of the Universe" and "Lady of Heaven." Known as the "Celestial Cow," she created cows, later becoming responsible for all farm animals, and punished farmers who neglected their livestock. She was also the protector of infants, perhaps because mothers and cattle both produce milk to suckle their young. Hathor embodied fertility and the Egyptians made her responsible for the female libido. They later associated her with love and sex and referred to her as "lady of the vulva." Unsurprisingly, the Greeks linked her with Aphrodite.

Hathor was also the goddess of dance and her cult encouraged the playing of music, with worshipers shaking the *sesheshet*, a sacred rattle, in her honor. Her priestesses also made music by shaking a ritual necklace made with heavy beads. Because she was the goddess of beauty, Hathor was often depicted on the back of women's mirrors. She was worshiped all over Egypt, and became closely identified with the great Egyptian goddess Isis.

Horus The falcon-headed god

Many boats sailing in the Mediterranean still have eyes painted on them, to bring luck and to help the sailor find his way. These eyes are directly related to the Eye of Horus.

Horus warded off evil spirits and in some areas of Egypt became identified as the morning sun, a manifestation of Re known as Re-Horakhety.

Horus is a solar god with the attributes of Re, and was depicted as a falcon or a falcon-headed man. He was often represented as falcon wings on either side of a solar disc; the pharaohs were considered to be the embodiment of Horus. In early traditions Horus and Seth were brothers. Later, when the myth of Osiris emerged, Horus was held to be his son and Seth his brother. It was as son of Osiris that Horus emerged to avenge his brutal murder by Seth. Horus lost an eye in his battle against Seth, later giving this eye to Osiris, a gift that helped Osiris to rise from the dead. The Eye of Horus became a very popular protection and charm, and the sun and moon were sometimes called the "Eyes of Horus."

Horus had four children, who protected the dead from thirst and hunger and guarded their entrails in the tomb.

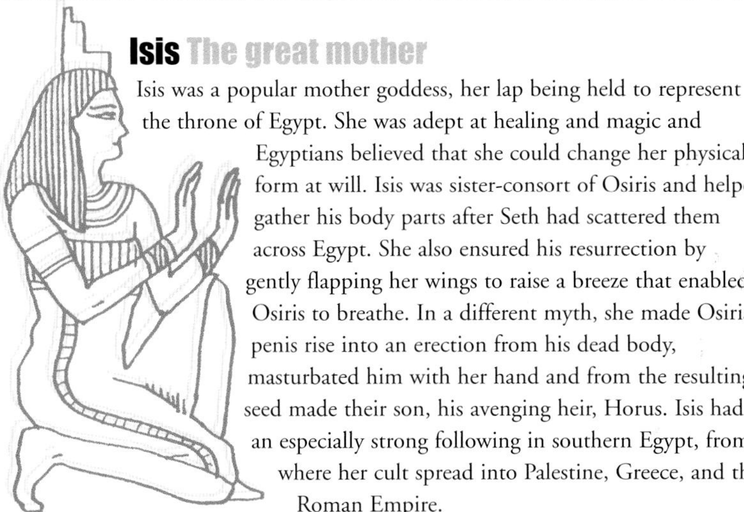

Isis The great mother

Isis was a popular mother goddess, her lap being held to represent the throne of Egypt. She was adept at healing and magic and Egyptians believed that she could change her physical form at will. Isis was sister-consort of Osiris and helped gather his body parts after Seth had scattered them across Egypt. She also ensured his resurrection by gently flapping her wings to raise a breeze that enabled Osiris to breathe. In a different myth, she made Osiris's penis rise into an erection from his dead body, masturbated him with her hand and from the resulting seed made their son, his avenging heir, Horus. Isis had an especially strong following in southern Egypt, from where her cult spread into Palestine, Greece, and the Roman Empire.

Isis was identified with a number of goddesses both in and out of Egypt. In Athens, she became Minerva; in Sicily, Prosperina; in Crete, Diana. Early images of the Virgin Mary with the baby Jesus are remarkably similar to drawings of Isis nursing Horus.

Khepri The scarab god

The Egyptians believed that the sun traveled through the underworld at night, and then up across the sky by day, before falling below the western horizon at dusk. Khepri, the sun god of dawn on the eastern horizon, being self-created and the creator of all life, who made the earth from his spit, was represented as a gigantic scarab (dung beetle), pushing the disc of the sun across the sky, like a scarab pushing its ball of dung. He was sometimes shown as a ram-headed beetle, or with a human body and a beetle-head. The scarab became a very common symbol in Egyptian art and frequently appeared on seals and protective amulets.

King Tutankhamun's tomb contained many semi-precious or gold scarabs, which connected Tutankhamun to Khepri and helped ensure his triumphant ascension to the heaven of the sun.

Khnum The divine potter

A god with the body of a man and the head of a ram, the creator god Khnum made the first man and woman on his potter's wheel, a deed that was also attributed to Ptah. When he became weary of creating life, he placed a potter's wheel in the womb of all females, giving them his creative powers. He was also a builder: he constructed a ladder that reached to heaven, and made ferryboats. Khnum was linked with the idea of fate for he gave an allotted time span to each person he created.

The flat-horned ram first associated with Khnum has now been extinct for three thousand years.

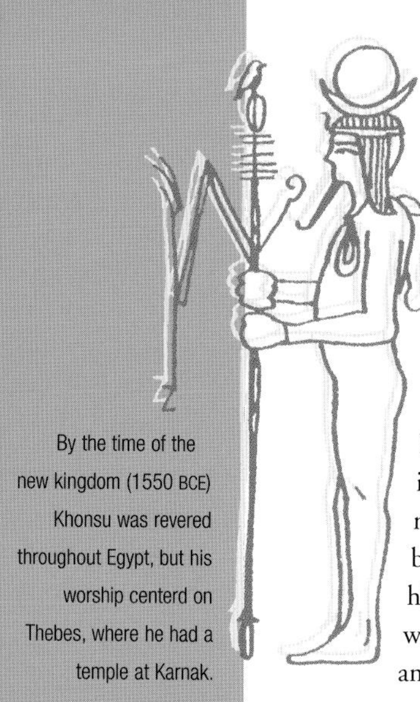

Khonsu The wanderer

Khonsu or Khons was initially represented as a vicious god who hunted, strangled and devoured the life force of other gods, feeding on their heads and hearts and taking their strength as his own. A thousand years later, especially at Thebes, he was seen as a much more benevolent figure and acquired the name "greatest god of the great gods." The son of Amun and Mut (although some traditions made him the offspring of Sobek and Hathor), Khonsu was a healer and exorcist. As the moon god and "wanderer" (his name referred to the journey he made each night across the sky), Khonsu was shown with the head of a hawk (a bird that could soar in the sky, just as Khonsu did). The lunar cycle helped the Egyptians mark the passage of time, so Khonsu was also the god of time. The baboon, which the Egyptians saw as a lunar animal, was sacred to him. Khonsu was sometimes depicted as a good-looking prince, wearing a lock of plaited hair over his right shoulder, as princes did, and carrying a scepter inscribed with the symbols for life and stability.

By the time of the new kingdom (1550 BCE) Khonsu was revered throughout Egypt, but his worship centerd on Thebes, where he had a temple at Karnak.

Maat The goddess of truth

The embodiment of truth and justice, the goddess Maat represented order and integrity in the universe. She prevented the forces of chaos from taking over the world, embodied the values that all mortals had to uphold, and protected the kings, who looked on her as being their authority to rule.

Sometimes represented simply by an upright ostrich feather, Maat was more commonly shown as a woman with a feather on her head. When Osiris judged the dead, the heart of the deceased was measured against the feather of truth, symbolizing Maat and a blameless life. When the dead were judged, Maat adorned the innocent with feathers to symbolize righteousness. Maat was the patron of judges, magistrates and all court officials.

When Egyptian judges passed verdict, they wore an image of Maat to show their authority and responsibility; she inspired them to act justly.

Min The firm one

As a fertility god, Min was one of Egypt's earliest deities. The Egyptians depicted him as a man standing upright, his legs bandaged like a mummy's. One hand was extended while the other hid under a cloak from which his massive, erect penis protruded. To further convey the image of rampant virility and fertility, the Egyptians gave Min black skin (the color associated with growth and resurrection).

He was sometimes portrayed as a falcon and worshiped in the guise of a sacred white bull (an animal associated in ancient Egypt with sexual potency). The Egyptians offered him a type of lettuce that had white sap, which they believed had powerful aphrodisiac qualities. Min was also a desert god. As such, he protected nomads, hunters, travelers seeking their fortunes, and migrant workers.

Min became the son of Isis and the father of Horus (somewhere between 2100 and 1600 BCE, before the cult of Osiris became dominant); he was later identified with the sun god Amun.

The Greeks linked Min with Pan and in the third century BCE re-named the city Khent-Min (Shrine of Min) as Panopolis.

Montu The warrior god

Montu emerged as god of war from around 2000 BCE. He embodied the conquering strength of the kings and in some traditions slew the demon-serpent Apophis. The warrior kings often added his name to their own and decorated royal warships with images of Montu striding aggressively towards the enemy, a threatening mace or spear in his hand. In ancient texts, he was shown as a man with a falcon's head topped by the solar disc, often winged. In some parts of Egypt, he was also linked with a type of bull called *bukhis*, with a distinctive white hide and black face; in this aspect, he was believed to cure various illnesses. *Bukhis* bulls were sometimes stabled at temples, where they were consulted as oracles and baited to fight with each other.

Montu lost his status as god of war when Egyptians changed the hierarchy and made him subordinate to the god Amun, although he continued to appeal to warrior kings.

Mut The divine mother

As vulture goddess and mother goddess, Mut was portrayed as a woman with the head of a vulture or as a vulture protectively covering her offspring. "The Great Mother" symbolized maternal love; she was also the mother of the Egyptian kings. The queens of Egypt adopted Mut's protective role and from around 1500 BCE onwards they wore a golden crown depicting vultures' wings, to indicate their protection of those they ruled. Mut was sometimes depicted as a lion, suggesting that she was approachable and could be stroked and pacified, but might also be dangerous and destructive. She was sometimes shown with a penis, perhaps to suggest a fiery temperament. Mut's cult had both priests and priestesses, who worshiped her as a divine female pharaoh. She became Amun's wife and chapels were dedicated to her in his temples.

Mut's name derives from the Egyptian word for "mother."

Neith The hunter

Late inscriptions reverse all the positive imagery of Neith, making her the divine mother of evil Apophis. Nevertheless, she was linked by the Greeks with Athena.

This goddess's name meant "that which is" or possibly "the terrifying." Originally a primeval bisexual creator goddess, Neith acquired several responsibilities in early Egyptian civilization. She was goddess of war and hunting, being symbolized by a shield and two crossed arrows. She is sometimes depicted carrying these attributes or wearing them as a crown. She was also portrayed with a loom above her head, for she was patron of weaving and blessed the funeral shrouds of the deceased. From around 1500 BCE, Neith became increasingly prominent. As the mother of the sun god Re, she effectively became the mother of every living being and many saw her as the creator of the world. She was also the mother of Sobek, the divine crocodile who guarded marshy waters.

Nekhbet Protector of Kings

Nekhbet was also symbolized as a serpent and snakes in the royal crown often referred to her power.

As a vulture goddess, Nekhbet was predominantly worshiped to the west of the Nile, where the vulture came to symbolize the region and also adorned the royal crest. Nekhbet was a protective goddess and the mythical mother of the king, whom she breast-fed. Her cult developed until she became responsible for the birth of all human babies. She was sometimes portrayed as a woman wearing the skin of a vulture on her head. When pictured as a vulture, she often clutched the symbols of eternity in her talons, perhaps to bestow divinity upon the king she protected. Wadjet, the snake goddess, was Nekhbet's companion. When depicted together, they were referred to as the "two ladies."

Nephthys The friend of the dead

There was no independent cult of Nephthys, but she served an important role, protecting the coffins of the dead and helping to ensure the safety of the jars containing the entrails of the embalmed. Since the Egyptians did not consider the soul to be completely independent of the body, in protecting the body Nephthys also helped guarantee the deceased an afterlife. She shared these responsibilities with her sister Isis and with Neith. Nepythys was wife to her brother Seth and sister to Osiris, and in the Osiris myth she helped Isis gather and protect Osiris's slain corpse. Often represented as a hawk, she mourned for the dead and stood behind Osiris in the great hall of truth where souls were judged. In one mythical tradition, Nephthys lusted after Osiris and visited him at night, when he confused her with his partner Isis. The result of their union was the god Anubis.

The name Nephthys meant "mistress of the mansion" and this title was often written on her head in Ancient Egyptian drawings.

Nun The void

Nun was the personification of the primeval waters from which all of creation sprang. Although worshiped as a god for many thousands of years, he did not engage with the human world. He represented the watery chaos out of which order would emerge, but which also continued to hold the potential for disorder. The foundation of all life, Nun was dark, mysterious, limitless. The creation of all that existed, including the gods, did not come from nothing, it arose from the void of Nun. In one central myth, creation began when a mound rose out of Nun, on which Atum gave birth to himself and then began to create others. In scenes depicting the act of creation, Nun appears as a man lifting the sun (itself carried on a boat) out of the primordial waters.

Late in the decline of the Egyptian empire, Nun became a symbol of negative power and disorder, being seen simply as an abyss. Early Egyptian Christians understood Nun to be a hell.

Nut The sky goddess

Nut was the goddess of Heaven and one of the very first Egyptian deities. Daughter of the air (Shu) and sister-consort of the earth (Geb), Nut bent over the earth, touching the horizons with her hands and feet. She was the mother of all the heavenly bodies, and their mistress. All that passed across the sky passed through her: each evening she swallowed Re and each morning gave birth to him, so the Egyptians saw dawn as the birth of the sun, and dusk as the sun being consumed by its mother so it could pass through her body. Coffins and tombs were painted with stars, representing heavens from which the deceased would emerge to a new dawn and new day, an afterlife, just as the sun emerged from Nut. The goddess was sometimes depicted on tomb walls as a naked woman, perhaps with vulture's wings, sometimes carrying a round container on her head. She was also represented as a giant cow.

Nut's body, arched in the sky, marked the very edge of the universe. Only formless chaos lay beyond.

Osiris The judge of the dead

Osiris's skin was often depicted as green (a color of rejuvenation), white (the shroud of the mummy) or black (symbolizing life after death).

Probably the best-known god in the Egyptian pantheon; Osiris's name meant "place of the eye" and he was symbolized by an eye. Later identified with the sun god Re, Osiris ruled the underworld, being called "the king of those who are not."

His jealous brother Seth (who wanted to inherit in his place) murdered Osiris and cut his body in pieces that he scattered all over Egypt. The dismembered body was collected and restored with the assistance of Osiris's son Horus, his sister Isis and Anubis. Isis breathed life into Osiris who, as a resurrected god went to the underworld, where he judged the dead and held the key to eternal life. A blissful afterlife used to be reserved for the kings and their closest companions, but the cult of Osiris extended the possibility of paradise to all mortals.

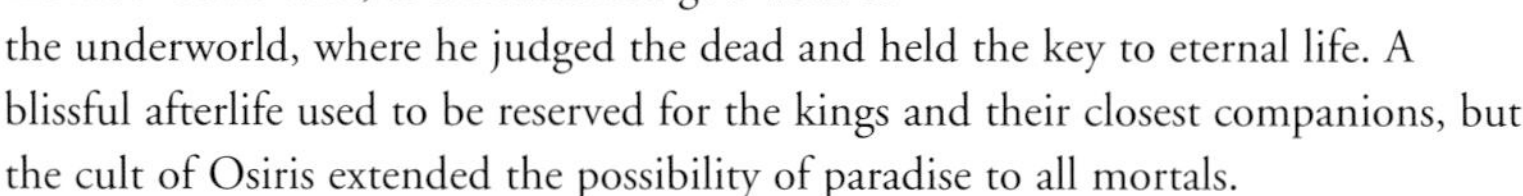

Osiris was also the god of fertility or nature. He was often depicted with his feet bound together like a mummy and holding crook – or whip-shaped scepters.

Ptah The god of crafts

The Greeks identified Ptah with their god Hephaestus – the ugliest of their gods.

The creator god who made the earth and the skies and molded the universe on a potter's wheel, Ptah spoke the word that began all things. He was the craftsman of the gods: he hammered out the floor of the heavens from an iron slab, he beat out the skies, made the stars out of crystal and ensured that the heavens were put in place. Known as "Lord of Truth" and "Great of Strength," Ptah molded royalty and shaped the destiny of mortals.

Ptah helped his son Osiris to survive in the underworld, giving him food and drink and magical weapons to destroy his enemies. The Egyptians begged Ptah, the "Hearer of Prayers" and protector of those in the afterlife, for care for the deceased and for help in this life.

He was depicted as a man shrouded in mummy wrappings, wearing a tight skullcap, carrying a staff in his right hand. His statues were often covered in gold leaf, with his scalp in a different color, most often lapis blue.

Ptah's cult was widespread throughout Ancient Egypt, with its center at the great temple complex at Karnak. Often equated with Atum, Ptah became one of the three state gods of Egypt, along with Amun and Re.

Re The sun god

Re, or Ra, was a sun god and creator god. He became the supreme, omniscient deity of the Egyptians, who called him "Glorious," "Strong," and "Mighty." When other gods grew in importance, they were often associated with Re (so Amun became Amun-Re, Horus became Re-Horakhty, and so on). Re emerged from the primeval chaos of Nun to create everything. By day, he voyaged across the sky in a solar bark and by night journeyed through the underworld, where Seth protected him from his enemies. In another explanation of the phenomenon of the sun rising and setting, the sky god Nut swallowed Re at dusk and gave birth to him at dawn. The sun was Re's central symbol. He was also depicted as a falcon wearing a solar disc, and sacred texts described him as an ageing king with gold flesh, silver bones, and lapis lazuli hair.

The early Egyptian kings identified themselves as Re: he was the source and proof of their authority, and ensured their journey after death to the heaven of the sun, where they joined him (and sometimes merged with him) to travel across the sky in his solar boat. He was closely related to the goddess Maat, whose Truth was important to his authority, and he fathered Hathor. The cult of Re was much more important to the kings than to ordinary people, for whom he was a remote figure, and they increasingly focused on Osiris in their daily lives.

Re's principal cultural center was at Heliopolis, near modern Cairo, which was named after his counterpart, the Greek god Helios.

Seth The desert god

Seth murdered Osiris and tried to kill his nephew Horus. He represented chaos and violence, but also strength. A trickster god, Seth became a hated and cursed figure within Egypt, although he did several good deeds, including subduing and conquering the evil serpent Apophis. When the king ruled efficiently, it meant that Horus and Seth were united.

By the first century BCE Seth was portrayed as an enemy of the gods – he was depicted him with an ass's head with a knife sticking out of it, the blade buried deep. He came to symbolize evil (especially after the cult of Osiris became popular) and was often cursed. Seth was associated with the desert (the opposite of Osiris's links with nature) and was also a god of the earth. When miners struck iron, it was the bones of Seth they had found, and when he breathed, worms came out of the ground.

Because Seth was associated with storms, especially those of the desert, the Greeks identified him as the legendary serpent-monster Typhon.

Shu The holder of the sky

Shu was chiefly worshiped at Leontopolis (Tel el-Muqdam) in the Nile Delta. In Greek, Leontopolis means "City of the Lion."

Shu was god of Air and Light (what we would nowadays call atmosphere). He fathered Geb (earth) and Nut (sky) and held up the heavens by kneeling on the earth and supporting Nut, the sky goddess, with his arms. Shu separated light from dark and was the partner of Tefnut; in the guise of lions they guarded the eastern and western horizons to ensure that the sun rose and fell. Mythically, Shu was a king of Egypt, who ruled for many generations until he became old and took his place with the other gods, attending the sun god Re.

Sobek God of crocodiles

The Greeks venerated Sobek to the extent that they named a town, Crocodilopolis, after him.

Depicted as a man with a crocodile's head, crowned with plumes, sometimes as a crocodile, Sobek was represented in some accounts as pre-dating creation and having laid the eggs of life in the primordial waters. He created not just crocodiles but all life on earth. His temple had a sacred lake that harbored a live crocodile, alluding to the origin of the cult of Sobek in marshy areas. Sobek was a benevolent deity, who could also be fierce and destructive; his name was often annexed to those of the kings to signify their association with him. He became identified with many gods, including Horus and Re, and at a later stage was seen as being increasingly involved in fighting and defeating Re's enemies.

Sokar God of goldsmiths

A god dating from prehistoric times, Sokar emerged in early Egyptian mythology as an agricultural deity. The Egyptians called him "cutter," as he worked the fertile earth alongside the Nile. He was also credited with making the bones of the kings and various objects that they would use in the afterlife. He later became associated with the perfumes and ointments used in Egyptian rituals. Sokar was assigned a more specialized function to help distinguish him from Ptah, also a crafts god, and became the god of goldsmiths. Nevertheless, it was presumably Ptah's priests who argued that Ptah had made Sokar and the two gods were closely linked: Sokar became Ptah-Sokar. By this stage, Sokar was portrayed as a man with a hawk's head, standing on a winged boat. Gradually another duty was attributed to Sokar: he helped to guard tombs. When Osiris emerged as lord of the underworld, Ptah-Sokar underwent a further change and became Sokar-Osiris. Later still he became Ptah-Sokar-Osiris.

During festivals, Sokar's followers wore onions around their necks to show their allegiance to him. Embalmers used onions in their work, so they were a suitable symbol for a god closely associated with the dead.

THE DIVINITY OF THE KINGS

The Ancient Egyptians believed that the kings or pharaohs, as they came to be known by around 1600 BCE, were divine. During their earthly lives, the kings took the title "son of Re" (the pharaohs were almost always male) and were often identified with the god Horus, who succeeded to the throne after his father Osiris, following a long conflict with Seth, a god of darkness. As Horus, the king could communicate with the gods.

During his coronation, and at major festivals, the king acted out Horus's victory over Seth. The ritual showed that the king brought *maat* (truth and order) to rule over the chaos that preceded his ascension to the throne. It was designed to demonstrate to all present that the king was the rightful leader of Egypt.

Tawaret Goddess of pregnancy

The hippopotamus goddess Tawaret was one of the ugliest deities in the Egyptian pantheon. She was depicted as part-woman, part-hippopotamus, standing upright with human arms, breasts and legs, an impressively round, pregnant-looking belly, a crocodile tail and a hippopotamus head. Tawaret's unusual features served a purpose. She assisted women during childbirth and, like the knife she held in her hand, her looks scared away the evil forces that might otherwise harm women in labor. Chiefly a domestic goddess, Tawaret ultimately took on a wider role, becoming known as "mistress of the horizon" as well as "great one." Tawaret was a very popular domestic deity, and her followers had her image on their beds and mirrors so that she would protect them.

Amulets of Tawaret continued to be made by the Romans in the early first century CE; they associated her with healing.

Tefnut Goddess of dew

In one mythic tradition, Tefnut created the morning dew from her vagina.

The lioness-headed goddess Tefnut wore a serpent and solar disc on her head, although she was more closely associated with the moon. She was the goddess of moisture or dew and the partner of her brother Shu (god of the atmosphere). Atum created her from his own body, by spitting her. On some tomb walls, Tefnut is symbolized by a set of lips producing spit.

One of the great cosmic gods of Egypt, Tefnut played a part in the judgement of the dead alongside Egypt's other major deities. In the Judgement Hall, she listened to the deceased perform the negative confession, where they denied that they had committed various evils and had their hearts weighed against Truth.

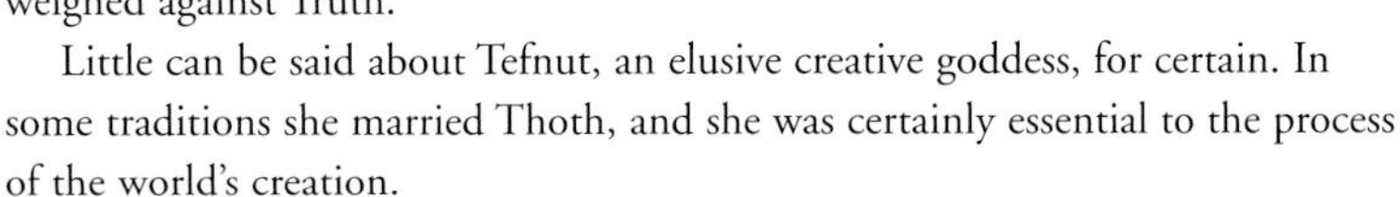

Little can be said about Tefnut, an elusive creative goddess, for certain. In some traditions she married Thoth, and she was certainly essential to the process of the world's creation.

Thoth God of wisdom

The Greeks equated Thoth with Hermes, who also assisted the dead, and renamed the city of Khmnu, center of Thoth's cult, Hermopolis.

As the god of wisdom and of magic, Thoth invented writing and was the patron of scribes, who formed an important profession in Egyptian society. He also acted as the scribe of the gods, often recording events like the Judgement of the Dead on papyrus. He was credited with authorship of many ancient sacred texts, including *The Book of Going Forth by Day.* Because he was a lunar deity, the Ancient Egyptians often depicted him with a lunar disc on his head, and he often appeared as a baboon, probably because he became the head of a baboon cult. He was also sometimes imagined as an ibis, a bird whose mighty wings could carry a king to the heaven of the sun. Thoth had many other duties: because the Egyptians used the lunar cycle to mark the passage of time, he became the god of time. He was also responsible for astronomy, mathematics accounting and languages.

In the Osiris myth, Thoth opposed Seth, protected Osiris and helped Isis resurrect him by showing her the spells she needed. As the helper of Osiris, he was also helper of the dead.

Wadjet The cobra goddess

The snake goddess Wadjet spat venomous fire at her enemies. Most commonly portrayed as a crowned cobra, she protected the Egyptian king, alongside the vulture goddess Nekhbet, although Wadjet's cult was based east of the Nile, unlike that of Nekhbet. Wadjet was also guardian of the god Horus, thus associating the king of Egypt with the god. The Egyptians believed that the sun god Ra wore the cobra emblem of Wadjet on his head (they sometimes drew him with her coils wrapped around the solar disk above him) and that she helped him fight the forces of evil that attacked Ra as he traveled through the underworld at night. Egyptian kings bore on their heads the image of Wadjet as a cobra, rearing up and poised to attack her enemies. This symbolized the king's legitimate authority and sovereignty, and Wadjet was held to appear in human form to crown the new king.

The sun disc or *re*, a powerful symbol in ancient Egypt, often had a cobra entwined around it, showing that Wadjet embraced the kingship of the sun god.

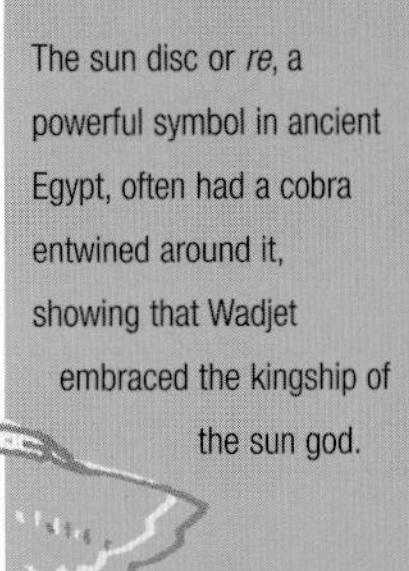

TEMPLES AND GODS

Egyptian temples symbolized divine order and were believed to be the mansions of the gods. The innermost sanctuary was the bedchamber of the god, and was surrounded by other bedrooms so that divine visitors could sleep. The priests were the servants of the gods and satisfied their heavenly needs with food offerings and sacrifices.

As such, the priests were the servants of the gods; they chanted spells, delivered offerings and also attended to the needs of mortals, both the living and the dead. Until late in Egypt's history, commoners were not allowed into the sacred center of the temple that housed the statues of the gods.

Wepwawet The war god

As "Opener of the Ways," the aggressive Wepwawet led the way into battle, holding weapons of war – a mace and a bow – in his hands. He was depicted as a gray – or white-headed jackal and was the champion of royalty, being closely linked to Horus and, by association, to all Egypt's kings. A destroyer of his enemies, Wepwawet was also the opener of the ways in death, for he led the funeral procession and conducted the dead through all the potential hazards of the underworld to a blessed after life. Closely linked with Anubis, the other jackal-headed god, Wepwawet watched over the tombs of the dead and also over the dawn, ensuring that it opened up the day.

The Greek name for his cult center, Lykopolis (wolf), shows that he was once or also worshiped as a wolf.

europe

Right: Balor, the Celtic-Irish god of death and king of the Fomorians – a race of demonic giants and ancient occupants of Ireland.

Below: A cromlech (remains of a Druid burial chamber) situated near Inverness, in the Highland region of Scotland.

The related myths of the Olympian gods of Greece and Rome, and of the Nordic and Celtic gods of western Europe form the warp and weft of contemporary culture. Although they may appear very different to modern readers, they did not develop in isolation from one another, or from the older traditions of Egypt, Persia, and Mesopotamia. From the earliest times, through trading contacts, the Celtic and Germanic peoples of northern and western Europe were in contact with the thought and beliefs of the Greeks and Phoenicians; the Greeks and Romans, through their far-flung conquests, absorbed many customs, gods and myths from the East and West, which they adapted to their own needs.

The Greeks had a sophisticated mythology and a multitude of gods, and by the time Greece became a province of Rome, in 146 CE, the Romans had already willingly long embraced the Greek divinities. They did so because although they had their own gods, they had a very limited mythology. Indeed, many of the Roman gods were anonymous: *Si deus si dea* (whether you be god or goddess) was a common way for them to be addressed. The Greek gods, on the other hand, had well formed characters and histories.

The Roman conquest provided the means for the gods of the Near East to enter European culture, and it also exposed the Romans to a new set of European gods. The Celts occupied large areas of Europe, and as the Romans expanded their empire, they assimilated many of the Celtic gods as Roman divinities. Celtic figures that had been worshiped locally became more widely recognized and took on Roman characteristics, and the Roman soldiers and imperial administrators of the Celtic provinces took these foreign gods and their mythology back to Rome.

By the time the Romans left Britain in the early fifth century CE, Christianity, with its one all-powerful god, was widespread. Yet it had not reached the northeastern borders of the Roman Empire where the Germanic peoples were already migrating into Scandinavia. As the Empire declined, the Germanic tribes conquered lands that had been under Roman occupation and also settled in Britain. They took their gods with them, and these divinities resurfaced centuries later in the Viking Age of the Nordic people, which lasted from the eighth to the eleventh centuries CE. Viking mythology resonates with earlier influences. The Norse goddess Freya, for example, is often linked with the Egyptian goddess Isis, and both Odin and Thor share many of the characteristics of Jupiter.

Top: Raphael's *Wedding of Cupid and Psyche.* In Roman mythology Psyche is a beautiful princess loved by Cupid, the god of love.

Bottom: Freya is the goddess of love and fertility in Norse mythology and also the symbol of sex and sensuality.

Above: The Romans invaded Greece in 146 CE and remained for several hundred years, adopting many Greek gods into their own pantheon.

GREEK AND ROMAN GODS

The ancient Greeks did not have a central sacred text like the Bible that recorded their beliefs about their gods. Our knowledge of the Greek deities comes mainly from Greek drama and the epic verse that was performed at public festivals. The most important early epics about the Olympian gods were composed around the eighth century BCE. Centuries before their conquest of Greece in 146 CE, the Romans had adopted many of the Greek gods to supplement their own rather meager, formless pantheon of divinities. But the Romans did not stop at importing the Greek gods; they adopted the gods and goddesses of many of the peoples they had conquered.

The poets Homer and Hesiod (c. 800 BCE) both described the relationships that the gods had with each other and with mortals. Unlike mortals, the gods did not grow old or die and the poems make clear that immortality was the unbridgeable gap between humans and gods. A favored few achieved immortality, such as the hero Heracles (Hercules), but this was the supreme gift of the gods, usually restricted to men of divine descent and to the leading figures of the "Heroic Age." Hesiod made a distinction between two different generations of gods, the Titans and the gods who lived on Mount Olympus under the leadership of Zeus. The giant Titans fought against the supreme god Zeus after he had ousted his father Cronus from power. Zeus emerged victorious to preside over all human and divine affairs and went on to punish many of the Titans. Prometheus was the son of a Titan and, in Hesiod's account, Prometheus's antagonism toward Zeus directly led to Zeus creating Pandora, the first woman, to punish mankind.

Although the Romans incorporated the Greek gods into their own religious system, they did not abandon their own gods. Several Roman gods had no Greek counterparts. The Romans deified Romulus, the mythical founder of Rome in 753 BCE, and they also worshiped the double-faced god of beginnings, Janus. When a Roman god became associated with a Greek counterpart, the Roman characteristics of the god remained: for example, the Roman god of war, Mars, reflected the military order, might and priorities of the Roman empire's army while his Greek counterpart, Ares, was wild, chaotic, and unruly.

Sometimes the Romans adopted a Greek god that was entirely new to their pantheon. Apollo, for instance, had no Roman equivalent. The Romans turned to this Greek god after two epidemics swept through Rome around 430 BCE. They worshiped him as a god of healing without changing his name to a Latin form, but they called him Phoebus in his role of the sun god, instead of the Greek Helios. Another key difference between the Roman and Greek religions was that the Romans adopted gods from all over the Empire. Through trade and conquest they encountered gods from other cultures, and they adopted several of them, including the Persian god Mithras and the Celtic goddess Epona.

Right: Hera, the wife of Zeus and queen of the gods, responded to her husband's frequent infidelities with anger, jealousy, and vindictiveness.

Aphrodite/Venus 'Born of the foam'

The goddess of beauty, love, and sexual desire, Aphrodite sprang fully-grown from the sea foam into which Cronus had thrown the castrated genitals of his father Uranus. An enchantress whose beauty filled the air with birdsong and made flowers bloom as she passed by, she wore a magic girdle that made both mortals and gods fall in love with her. Known by the Ancient Greeks as "the unholy" because of her habit of destroying mortals who neglected her, Aphrodite was an essentially flirtatious and uncaring goddess who could drive those who looked at her to a frenzy of desire. Only the virgins of Olympus – Artemis, Athena, and Hestia – were immune to her love spells. She married Hephaestus, the ugliest of the gods, who trapped her in bed with Ares, much to the amusement of the other Olympian gods. She and the goddess Persephone both fell in love with the handsome Adonis, who shared his time between them.

At Aphrodite's temples at Corinth the priestesses were prostitutes. Aphrodite protected courtesans, and was the goddess of all the arts of lovemaking and beauty, including the uses of cosmetics and love potions.

Apollo The shining one

The smooth-cheeked, handsome and eternally young Apollo had many responsibilities: he was the god of prophecy, archery, music, and medicine; he also inspired harmony both on earth and on Olympus. Equipped with a lyre and a bow with deadly arrows, he could calm strife and destroy evil. He was a son of Zeus and twin brother of Artemis, later being identified with Helios the sun god because of his brilliance. Apollo was born on Delos, an island that became sacred to him, and one of his first achievements was to slay a python sent by the vengeful Hera to kill his mother (Leto) when she found that her husband had made Leto pregnant. He successfully competed against other deities for the role of oracle, became the god of music after beating Pan in a music competition, and led the Muses in their activities. Although Apollo was a rational god, representing civilized order and associated with sunlight, he could be ruthless when angered: his face would appear black as night. He had the satyr Marsyas flayed alive for challenging him to a music competition and gave King Midas ass's ears for preferring the music of Pan.

The modern phrase "a perfect Apollo" refers to a figure of youthful male beauty.

Ares/Mars The god of war

Romans identified Ares with Mars, a figure held in much higher regard in the Roman pantheon, second only to Jupiter.

To the ancient Greeks, the god of war was an odious figure who revelled in conflict, enjoyed slaughter, and gloated over the devastation he caused. Wild and headstrong, he rarely acted with dignity or any consideration for the legitimacy of his actions or their consequences. Son of Zeus and Hera, he embodied the brutality of war and the frenzy of fight. The poet Homer called him "the bane of mortals," and "a blood-stained stormer of walls." In the *Iliad* he was wounded in battle by Diomedes, a mortal fighting with the assistance of Athena who, unlike Ares, used reason as part of her weaponry. Ares was the lover of Aphrodite, and appeared in Greek art as a bearded warrior and as a naked young man; his symbols were a spear and a burning torch. To the more militaristic Romans, Mars was second only to Jupiter in power. He was the father and protector of Romulus and Remus, the founders of Rome. As the god of strength he guarded Rome's borders. He was often shown in full armour and was a favorite with the soldiers of the empire.

Artemis/Diana The virgin huntress

Artemis received human sacrifices. On the coast of Taurus and at Attaca, shipwrecked sailors were ritually killed in her honor.

Apollo's twin sister, Artemis, was the Greek goddess of hunting, widely worshiped as the virgin deity who lived in the "wild country." An earth goddess who hunted daily in a chariot drawn by stags with golden antlers, she was the mistress and protector of animals. In neglecting their livestock, mortals offended the goddess. Artemis also enjoyed dancing and singing and the company of a retinue of nymphs. In Greek art, she was often pictured flanked by the wild beasts she hunted. Artemis presided over childbirth, childrearing and the key stages of female development, as well as the male rites of passage to adulthood. She brought prosperity and long life to mortals she favored, but she could punish those who displeased her, sending plague and misfortune. Her priests and priestesses were celibate and on Olympus Artemis was impervious to the advances of various gods.

Athena/Minerva Goddess of wisdom

The virgin goddess of wisdom, Athena maintained justice in society, especially in Athens, the city she protected. An impressive battle strategist and fearless warrior, she was known as the defender of city walls, often depicted wearing a helmet and carrying shield and sword. She was also the goddess of weaving (which she taught to Pandora, the first woman) and patron of learning, of art and of science. Athena had prophetic powers, which she could grant to mortals along with youth and majesty. The Romans worshiped her as the patron of business, industry, and schools. They portrayed her fully armed with a spear and helmet.

Zeus called Athena "dear gray-eyes" and looked on her as his favorite child. Her birth showed how close she was to Zeus: when he complained of a headache, Hephaestus split his head open and Athena jumped out, fully formed.

Athena was the guardian goddess of Athens, and the only deity to have a city named after her. The Parthenon was built in her honor. The Romans worshiped her as the patron of business, industry, and schools.

Atlas God of knowledge

Zeus punished Atlas the Titan, one of the primeval gods and brother of Prometheus, for taking sides against him in the civil war of the gods; he made him carry the sky on his shoulders. This simple punishment had a subtle elegance. Atlas was said to have invented astronomy, metaphorically carrying the sky in his mind, as knowledge. His punishment transformed knowledge into a burden: living at the edge of the world, he had to support the very sky that had once been his intellectual focus. He managed to relinquish his load only once, when Heracles took it upon his own shoulders while Atlas stole for him Hera's sacred apples from the garden of Hesperides, which Heracles needed to fulfil one of his twelve labors. Atlas had no intention of ever carrying the sky on his shoulders again, but he was tricked into resuming his burden temporarily while Heracles made himself more comfortable. Atlas was then left to endure his eternal punishment.

The Greeks named the Atlas mountains in Morocco after Atlas, believing them to be high enough to support the sky.

Chaos The void

The word "Chaos," which the Greeks used for this primeval god, remains in common usage as a term denoting disorder and confusion.

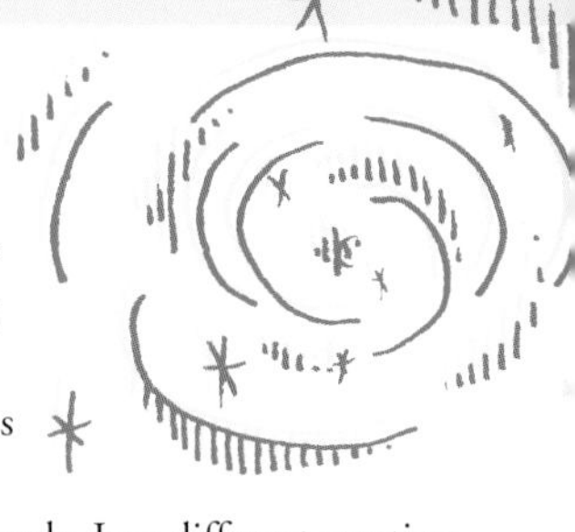

The gaping void, the potential of all matter and being: according to the Greek poet Hesiod, Chaos was a poorly defined primeval figure, personifying the primordial element from which the universe emerged. The Roman writer Ovid described Chaos as an unfashioned lump from which came the Earth, the sky, the stars, the winds, and the other gods. In a different creation account, Cronus was the first god, with Chaos following later and siring an egg that hatched into Phanes, a hermaphrodite creator figure.

Circe The sorceress

The story of Circe was a popular subject in Western art to remind the viewer of the ignorant pleasures that can turn men into animals.

The enchantress daughter of Helios, Circe combined powerful magic with a beautiful voice and often sang to her loom. She could be generous but was also prone to jealousy and had little love for mortals. Her palace was in a clearing on the middle of her island, where she lived like a queen, celebrated for her knowledge of herbs and magic. The immortal sorceress charmed unsuspecting visitors and beguiled them with her beauty. When they drank from the enchanted cup she offered, they turned into swine she then penned in a sty. They had good cause to bemoan their desire for a wine that made beasts of them. In Homer's *Odyssey*, Circe detained Odysseus for a year as her lover, after Hermes had protected the hero from Circe's magic with a plant of his own.

Cronus The bravest of the Titans

Although the gods warred, the Greek poet Hesiod stated that when Cronus ruled in heaven, mortals enjoyed a blissful golden age and immortality on earth.

Cronus, the youngest and bravest of the Titans, was the son of Uranus and Gaia. Uranus hated and feared his offspring and forced his children to remain within the womb of their mother. On Gaia's urging, Cronus castrated Uranus with a golden sickle and threw the useless organ into the sea, from which Aphrodite emerged. Cronus freed his siblings, but in an act that harked back to his relationship with his father, he later swallowed his own children to prevent them from overthrowing him. Their mother, Rhea, protected Zeus who grew up to defeat Cronus in a civil war, sending him to the grim underworld of Tartarus.

Demeter/Ceres The goddess of corn

Demeter, the goddess of vegetation and fruitfulness, was a tall, fragrant, beautiful figure who blessed those she favored with a good crop, but whose displeasure brought all the hardships of famine. The sister of Zeus and the daughter of Rhea and Cronus, Demeter was the third generation of nature deities, her mother's mother being Gaia. She consorted with both Zeus and Poseidon and was so close to Persephone, her daughter by Zeus, that they are often viewed as being two aspects of the same goddess. When Hades abducted Persephone and took her to be his queen in the grim underworld, Demeter grieved. Torch in hand, she roamed the earth looking for her daughter. In her grief and anger she avoided Olympus because Zeus had consented to Persephone's abduction and the earth withered into famine until the sacrifices to the gods were themselves in jeopardy. Zeus finally intervened and Demeter was reunited with Persephone for the summer months. Her mood improved and she restored the fertility of the earth.

At Eleusis, Demeter taught mortals the mysteries which would ensure initiates prosperity in this world and a blessed afterlife. She did so because the inhabitants of Eleusis offered her hospitality when she searched for her daughter. Demeter was especially worshiped in Athens, and was also linked to the Egyptian goddess Isis.

Dionysus/Bacchus "Twice-mothered"

The god of wine, ecstasy, and vegetation, Dionysus discovered the grapevine and the making of wine. His father was Zeus, who impregnated the Theban princess Semele, but she died through the machinations of Zeus's angry and jealous wife Hera, before she could give birth.

Zeus saved Dionysus by sewing his unborn son into his own thigh until the child was ready to be born. Dionysus journeyed to the underworld and rescued his mother, who became a goddess on Olympus. As a punishment, Hera drove Dionysus mad. He wandered the earth in a state of enchanted madness, until the goddess Rhea cured him and taught him the mysteries that would ensure a happy afterlife for his initiates. Accompanied by a retinue of frenzied followers who cried out and clashed their cymbals in his honor, Dionysus continued to roam, punishing any mortal who doubted his divinity. When the Thebans refused to worship him, he turned them murderous and mad.

Dionysus was often depicted as a bull-horned figure, suggesting virility, and his worshippers ritually carried huge wooden phalli on his festivals. Nevertheless, unlike Pan, he was not drawn with an erection.

Fortuna The goddess of fortune

The idea of the wheel of fortune stems from Fortuna; the Romans envisaged her standing on a wheel.

Fortuna originated as a fertility goddess and became a goddess of fate and luck. Roman emperors kept her statue in their bedrooms. She had many shrines in Rome where her cult was open to all. Fortuna was particularly popular with the poor, who formed part of the crowd at her festival when, according to the writer Ovid, excited devotees crowded along the river banks or into small boats to witness sacrifices to her. Fortuna was portrayed holding a boat rudder, to show that she guided human destiny, but she was also depicted as blind, suggesting that the decisions of the goddess were random. That image is at odds with another of her responsibilities: she helped women in childbirth, possibly because she was herself the first-born of Jupiter.

Erinyes The angry ones

According to the writer Aeschylus, the Furies haunted Orestes – the son of Agamemnon – for killing his mother Clytemnestra (even though she'd been adulterous to his father, whom she herself had killed). Originally his matricide brought no blame, as she was a despicable figure, but the Furies pursued him and he was tried for the crime because he killed without authority. Athena gave the casting vote, freeing him, and placated the Furies.

Better known as the Furies, this terrifying trio sprang from the earth when it was made pregnant by Uranus's blood after his castration. They attended Nemesis and acted as specters of vengeance. Grotesque figures with thin ghostly faces and flaming or blood-shot eyes, they had snakes instead of hair, which lashed and bit them, driving them to frenzy. They relentlessly pursued wrongdoers both day and night, especially those who had committed blood crimes within the family. They could also blight offending nations with war, pestilence, and famine. Terror, rage, and death followed in their wake. They were the pitiless executors of curses (especially those of a parent) and upholders of the natural order of inheritance and succession. The Furies – Tisiphone, Alecto and Megaera – were placed by Homer in the underworld, where they lived until a fresh curse brought them into action. They were eventually placated by Athena, after which they were re-named the Eumenides, "soothed ones."

Eros/Cupid God of desire

The Greeks had two gods called Eros. The first came into being soon after creation, and represented the cosmic principle of love. According to Hesiod he was the most beautiful of the gods, a force that ordered the universe. The second Eros – our Eros – was the son of Aphrodite and (usually) Ares, and a much more playful character who capriciously created sensual love and could be both cruel and cunning. The love darts he shot from his bow never missed: whomever they hit fell in love. He also had blunt leaden arrows that provoked aversion. Eros himself fell in love with the mortal Psyche but abandoned her when she turned to look at him. To show that he acted blindly, he was often depicted with his eyes covered and sometimes with golden wings.

The Roman god Amor, renamed Cupid by the Latin poets, was the Roman version of Eros, and is central to St Valentine's Day traditions.

Gaia Mother Earth

Gaia or "earth" emerged after Chaos and gave birth to Uranus, whose castration she orchestrated after he forced Cronus and their other children to remain inside her. She was the Great Mother, the power that sustained all life, later equated with Demeter. Gaia has enjoyed a renaissance in New Age and scientific communities since the 1960s, when the scientist James Lovelock developed his Gaia Theory, his hypothesis that the planet earth is a self-regulating super-organism. New Age followers of Gaia believe the Earth Mother is a guardian of the eco-system.

There was an important oracle at the ancient sacred site Delphi, in Gaia's honor; she was sometimes depicted as a woman emerging from the ground.

Graces, The Daughters of joy

Also known as the Charities, the three Graces, daughters of Zeus and the Oceanid Eurynome, were fertility figures personified as charming young women named Aglaia (joy or brilliance), Euphrosyne, (mirthful) and Thalia, (flowering). Their companions were the Muses, and like them they lived at the foot of Mount Olympus.

The Roman philosopher Seneca explained why the Graces were depicted in art naked, young, and holding hands. They were naked because kindnesses should be given sincerely and openly, young because the memory of a kindness should not be allowed to grow old, and holding hands because of the reciprocal nature of friendship.

The ancient Greeks used to advise melancholic or austere people to sacrifice to the Graces.

Hades/Pluto God of the underworld

Hades was the ruler of the underworld, a dread place for both the Greeks and Romans, for whom the afterlife was a desolate and unhappy existence.

The brother of Zeus and Poseidon, Hades ruled the underworld, a dark and miserable place where he lived in a splendid palace, built by Hephaestus, using the labor of the Titans. The hound Cerberus guarded the gates of his kingdom, which also came to be called Hades, and the souls of the dead crossed to it with the aid of the miserly and morose ferryman Charon. For Homer, Hades was the god of the dead, but he evolved into the god of death and a consumer of corpses. Hades wanted to marry Persephone, daughter of Demeter, but the goddesses of Olympus refused – his appearance mirrored his home – so he abducted her. As no mortal who ate in the underground could leave, a rule that Hades extended to the gods, he tricked Persephone into eating the seeds of a pomegranate. Zeus decided that Persephone would have to divide her time and live for the winter months with Hades. The Greeks depicted him as a majestic old man holding a sceptre and sometimes a pomegranate.

Hecate Goddess of magic

Occult groups in Thessaly worshiped Hecate. They dressed in clothes the color of the midnight sky and sacrificed lambs to her, letting the blood seep into the earth – as Uranus's mythically did - and invoking the goddess. They also summoned ghosts who would whisper the incantations that the worshipers wished to learn.

The cousin of Artemis, and often regarded as another aspect of her, Hecate was goddess of magic and a key underworld figure. There she helped guard the gates to Hades. Mortals could not bear to look at her frightening figure; she had serpents for feet and snakes for hair. The word Hecate meant "distant one," and the Greeks associated Hecate with the moon. As a figure of the night, she lingered on tombs or where the blood of murdered mortals had been spilled. She was accompanied by ghosts and demons and howling dogs. Hecate was the goddess of the paths of the night, and their crossroads. She was invoked by witches as the patron of witchcraft, and her followers offered her candle-lit cakes and scavenger fish. She could ensure prosperity and good fortune, however, and favoured the young.

Helios/Phoebus The charioteer of the sun

Later identified with Apollo, the god of light, Helios personified the sun. Daily he drove across the sky in a golden chariot made for him by Hephaestus, led by winged white horses that breathed flames. At dusk, when he descended in the west he would sleep in a golden boat (another creation of Hephaestus) until dawn, when he would rise again. As the god of light, looking down from the sky, Helios could witness everything that took place on earth and much that transpired in heaven. He alerted Demeter to the abduction of Persephone and told Hephaestus that Aphrodite was having an affair with Ares. It is perhaps because of this honesty that he became the witness of oaths.

The father of Circe and Phaeton, Helios had an especially close relationship with the island of Rhodes. When Zeus initially gave parts of the world to the gods as gifts Helios, being high in the sky, was absent from the gathering and received nothing. When he later saw the island rising from the sea, he claimed it for himself.

The statue of Helios made by the sculptor Chares on the island of Rhodes was one of the wonders of the ancient world. It was so large that ships passed between the legs.

Hephaestus/Vulcan The divine blacksmith

Hephaestus was the divine smith and architect of the gods whose impressive workmanship facilitated their lives. He built the palaces of the Olympian deities out of brass, designed tables and chairs that could move themselves in and out of Zeus's great hall as necessary, made lethal weapons, armour that no arrow could pierce, and designed jewelery for the goddesses. He also made golden shoes for the Olympians with which they could walk on air or water. Since they also traveled by chariot, Hephaestus shod the celestial horses' hooves so they too could travel through the air and over water.

The son of Zeus and Hera, Haephaestus was born crippled and one tradition holds that Hera threw him off Mount Olympus when he was born, in disgust. In another account, his father Zeus threw him down to earth for taking Hera's side in a row between his parents. Yet Hephaestus returned both to Olympus and back into his father's favor by acting as Zeus's midwife, splitting Zeus's head open so that Athena could emerge.

Hephaestus, the ugliest of the gods, married the beautiful Aphrodite, who did not remain faithful. He avenged her infidelity with Ares by trapping them both in a net, much to the amusement of the other gods.

Hera/Juno The goddess of marriage

The Greeks usually portrayed Hera wearing a crown and accompanied by a peacock, which was her symbol. She often held a sceptre on which a cuckoo perched, a reference to the mythical account that Zeus once turned himself into a cuckoo to seduce her.

Hera originated as a sky god and probably pre-dated the Olympian pantheon. As the wife of Zeus, she was Queen of Olympus and sat enthroned beside her husband. She diligently maintained her beauty by bathing in magical waters each year, using an oil that scented the world. Nevertheless this did not prevent Zeus from committing countless infidelities. Hera responded with jealousy and a long-term scheming vindictiveness. After Zeus fathered Heracles by a mortal mother, Hera first sent snakes to kill him and then later drove him to a madness in which he killed his wife and children. When she did show kindness to someone, it was often because of her hatred for someone else. She punished anyone who obstructed her. The Greeks worshiped Hera as the goddess of marriage and childbirth, presumably because of all her efforts to maintain her marriage. Zeus once became so infuriated with her that he bound her in chains and hung her from heaven.

Hermes/Mercury "Leader of souls"

The Greek Hermes became Mercury to the Romans. Later, northern Europeans identified Odin with Mercury, because Odin also had responsibilities for the souls of the dead. Like Hermes, the Roman Mercury used a caduceus, but Odin did not.

As the messenger of the gods, Hermes acted as ambassador and herald, communicating the wishes of the gods to mortals, and often serving as an intermediary between the gods themselves. He had strong magical powers and preferred negotiation to violence. To ensure that his envoy acted with speed, his father Zeus gave Hermes a winged silver cap and silver-winged sandals. Hermes also escorted the dead to the underworld and was seen as a god of communication, having invented the alphabet. He carried a powerful golden wand, or caduceus, a gift from his half-brother Apollo. It had the power to unite those who were divided by conflict. Hermes tested the wand on warring snakes, which immediately clasped each other in an embrace and wrapped around the caduceus. Travelers, merchants, and thieves all worshiped him as a bringer of good fortune. Hermes fathered the musical god Pan and also invented the lyre.

Hestia/Vesta Goddess of the hearth

The oldest daughter of Cronus and Rhea, favourite sister to Zeus, Hestia refused to marry Poseidon and Apollo and remained a virgin. As goddess of the hearth, she was depicted in art as a veiled young woman but does not feature in Greek myth because she always remained by the household fire. Traditionally Greek maidens were responsible for maintaining the hearth where Hestia received offerings of food and drink. In Roman mythology Hestia assumed a more significant role as the goddess Vesta. Oil, wine and water were all sacred to her.

Because Hestia was a virgin, the Greeks sacrificed to her cattle that were less than a year old and had therefore not engaged in any sexual activity. The first fruit of the harvest was offered to her.

EMPERORS AND GODS

Julius Caesar was the first Roman ruler to be deified after his death. First, he himself claimed divine descent from Mars and Venus, and the Senate (Rome's ruling body) confirmed his status as a demi-god. In 44 BCE a statue of him appeared in Rome with the inscription, "To the Unconquered God," and temples and altars were dedicated to him. Shortly after his murder the same year, he was formally declared a god of Rome.

Caesar's successor, Augustus, did not allow the Romans to consider him as a god, fearing to offend them, but encouraged the Egyptians to worship him (they already considered their own Pharaohs to be gods) and was happy to be called *Divi filius*, "the son of a god." At Augustus's cremation, an eagle was released from his funeral pyre to rise into the sky as a sign that his soul was heading to a destiny of immortal glory in the heavens.

Janus The god of beginnings

The Roman two-faced Janus had no Greek counterpart and was adopted by no other culture or mythology. He was able to look forward and backward at the same time and could see the future as clearly as the past. In a domestic context, he was the god of doors and gateways. His faces enabled him to look on either side of the door, to the outside and the inside, to the departure and the return. Doorways are a kind of beginning, one walks through the door and into a journey.

The Romans linked Janus to the journey, making him the god of harbors and the mythical inventor of navigation. Janus had a cult statue that stood in the center of the Forum, the chief meeting-place in Rome and the site of the Senate. In times of war, the doors of his double-gated temple were open and they were closed again in times of peace. Rome's administrative calendar began in January – a month named after the god – and the first day of January was sacred to him.

The Romans frequently stamped images of two-faced Janus on their coins.

Muses Goddesses of inspiration

The concept of a muse as a figure of creative inspiration continues to thrive in the arts.

The Muses were the divine offspring of Zeus and Memory (Mnemosyne) and lived at the foot of Olympus. The nine sisters were carefree and beautiful young virgins, figures of inspiration and entertainment. They presided over the arts, danced and sang, played musical instruments, and had prophetic powers. Each Muse had a different responsibility. Calliope was the muse of epic poetry, Clio of history, Erato of amorous and lyric poetry, Euterpe of music, Melpomene of the art of tragedy, Polyhymnia of rhetoric, Terpischore of dancing, and Thalia was the muse of comedy. Urania was the muse of both mathematics and astronomy. In Greek and Renaissance art they are depicted with the symbols of their relevant skills. Euterpe, for example, carried a flute and wore a crown of flowers, while Urania had a tiara of stars, a robe of heavenly blue, held a glove in her hand and was surrounded by scientific instruments.

ROMAN WORSHIP OF THE SUN

The cult of Sol Invictus, the Unconquerable Sun, flourished in Rome from the time of the Emperor Nero's rule (54–68 CE) onward. Nero himself adopted the symbol of the rays of the sun on his crown, to show his sovereignty as an incarnation of Mithras. A monotheistic form of Mithraism, Sol Invictus celebrated the rebirth of the sun each year on 25 December, at the height of the Saturnalia festival. When Emperor Constantine (who ruled from 324 to 337 CE) forbade the persecution of all forms of monotheism, and in so doing enabled Christianity to prosper in safety, he did so as the chief priest of Sol Invictus.

Mithras God of light

Worshippers of Mithras regarded him as "protector of the empire."

Mithras, an early Persian god, killed a wild bull in a divine battle. The blood from the bull's slit throat leaked to the ground, out of which sprang plant life. He was adopted by soldiers and administrators of the Roman empire, who worshiped him as Mithras. In the first and second centuries CE Mithras was particularly associated with the importance and value of loyalty, and was also linked to the sun, as a god of light. The cult of Mithras, which provided an alternative to worship within the official religious system, was exclusively male and did not extend to the Roman ruling elite. Worshippers met underground in caves, often decorated with signs of the zodiac, where they sacrificed bulls in his honor as part of a mystery cult. Initiates hoped to learn Mithras's secret knowledge, his mysteries, which would enable them to enjoy a blessed afterlife.

Oceanus The ocean

The first of the Titans, Oceanus personified the Ocean, the great river that surrounded the earth, on the far side of which lay the paradisiacal islands such as the Hesperides and the Isles of the Blessed. With his consort-sister Tethys, he had 3,000 sons (the rivers on earth). Oceanus invigorated the stars: at the end of night, when the stars fell, they dipped into the waters of Oceanus, and were re-energized for their journey the following evening.

When Poseidon emerged as the god of the sea and rivers, Oceanus effectively took retirement from the Greek pantheon and went to live at the edge of the earth.

MOUNT OLYMPUS

Olympus is the highest mountain in Greece, and the ancients believed that its peak touched the sky. According to the poet Homer, if one fell from Olympus it would take nine days and nights to reach the earth.

The gods lived at its summit and mortals were never allowed to visit their eternal home. To prevent them from trespassing, Olympus was wreathed in clouds that acted as gates, which were diligently guarded by the Seasons.

Zeus, the ruler of heaven, regularly held court at his Olympian palace. The gods traveled there by riding their chariots along the Milky Way. When the gods met, they feasted on ambrosia and nectar (the food and drink of the gods) as they listened to Apollo playing his lyre and the Muses make their music. Occasionally, they disturbed their idyllic retreat by arguing with one another and becoming involved in all kinds of rivalries.

Not all the gods lived on Olympus: Pan, for example, made his home in the region of Arcady, and Circe lived on her island. It was, however, home to the so-called Olympian gods, twelve deities considered extremely important in Greek culture. Those twelve were conventionally: Zeus, Hera, Athena, Apollo, Artemis, Ares, Hephaestus, Hebe, Aphrodite, Hermes, Hestia, and Demeter.

Pan/Faunus The shepherd god

A rural deity and fertility figure who presided over nature, Pan was the god of shepherds, hunters, and fishermen. The son of Hermes, he had the torso and head of a man, with horns on his head, dark skin and a long beard. Below that, he had the legs of a goat; his appearance alluded to the wildness of nature. Essentially a playful god, Pan was easily angered and could inspire fear. The Greeks thought strange noises in woods and glades were the sounds of Pan at play. He invented the reed pipe after failing to seduce the nymph Syrinx. She resisted his advances, appealing to the stream deities to rescue her; when Pan tried to embrace her, they turned Syrinx into reeds. Pan liked their soothing sound, and made them into a pipe.

Pan lived in Arcady, an agriculturally poor mountainous region of southern Greece. The Roman poets, especially Virgil and Ovid, transformed his home into the mythical Arcadia. For Ovid it was the home of primeval savages, for Virgil a paradise of groves and meadows, permanent spring and easy love.

Persephone Queen of the dead

The myth of Persephone's abduction was performed as sacred drama at Eleusis, where worshippers ritually searched for her with torches.

Demeter's daughter Persephone was originally Kore, a joyful goddess of youth who presided over the nymphs. As she picked flowers, the earth suddenly opened and Hades, who had fallen in love with her, rode out on a chariot drawn by black horses to seize her. He abducted her to his underworld kingdom, where she was called Persephone. Because of her mother Demeter's rage and grief, Hermes was sent to retrieve Persephone, although Zeus had promised her to Hades. But Hades had persuaded her to eat the seeds of a pomegranate and in tasting it she had tasted of death. Her link to the underworld became permanent; she had to spend the winter months in his underworld kingdom, but could enjoy the summer months reunited with her mother. As queen of the dead, who helped to determine their fate, she was a figure of fear.

Poseidon/Neptune God of the seas

The Romans identified Poseidon as Neptune, a water god and patron of water travel.

Poseidon became god of the seas when his brother Zeus took power from their father Cronus. He was also the god of earthquakes (Homer called him "earth shaker") and the god of horses.

The Greeks pictured him as a bearded old man with wavy locks carrying a trident (a large three-pronged fork) with which he could divide, calm or stir the waters. He could cause or calm storms at will and glided across the waves in a pearl-colored seashell chariot drawn by golden sea horses, with dolphins playing alongside. Poseidon married the minor sea deity Amphritrite and through her sired numerous sea creatures. He also mated with some horrendous figures, including the monstrous gorgon Medusa (who turned mortals who looked at her to stone) and one of the Furies. Through Medusa he fathered the winged horse Pegasus. He disguised himself as a stallion to mate with Demeter, who had turned herself into a mare to avoid him. This union resulted in the birth of the swift horse Arion. All horses were sacred to Poseidon; he created them and he taught mortals how to ride them. At Argos, his followers sacrificed horses to him by throwing them into a whirlpool.

Prometheus The creator of mankind

As the son of the Titan Iapetus, Prometheus was by nature ill-disposed to Zeus. Iapetus had sided with Cronus to fight against Zeus, and as a result endured the grim torments of Tartarus. Prometheus, a cunning trickster figure, opposed Zeus in a more subtle fashion; he fooled him into choosing the least tempting part of a sacrifice instead of the more satisfying meat. This angered Zeus, who punished Prometheus by withholding fire from mankind. However Prometheus stole fire from the chariot of the sun and gave it to mankind. In revenge, Zeus had him chained to a rock, where an eagle daily plucked at his liver, which each night would be restored. Prometheus was finally released by Heracles, to whom he gave directions to the distant Hesperides. Prometheus's name meant "forethought" and he could see the future. He set himself in opposition to Zeus by giving mankind fire, which differentiated them from the animals; he also gave them blind hope to enable them to endure the knowledge that they must die. In one mythical tradition, Prometheus created men from clay, whereupon Zeus ordered the creation of the first woman, Pandora, to punish mankind for having acquired fire. In Hesiod's misogynistic account, the alluring Pandora had "a bitch's mind" and "a knavish nature." She inadvertently opened the jar that contained the leftovers from creation and in doing so released all the ills of the world.

Prometheus enjoyed a renaissance with poets and musicians during the nineteenth century. The poet Percy Shelley's "Prometheus Unbound" hails his defiance of an oppressive god. Mary Shelley's *Frankenstein*, about a man who created a monster, is subtitled "The Modern Prometheus."

Rhea The mother of Zeus

The Titan Rhea married her brother Cronus and gave birth to Hestia, Demeter, Hera, Hades, Poseidon, and Zeus. Fearing that his children would usurp him, Cronus swallowed them whole at birth. Rhea saved Zeus from this fate by giving Cronus a stone disguised as a baby to swallow instead. She then hid Zeus on the island of Crete. Later Cronus vomited up the rest of his children, and Zeus led them in a war against the Titans, whom they defeated. The Greeks depicted the earth goddess Rhea sitting crowned upon a throne or riding a chariot pulled by lions.

At the sacred temple of Delphi, priestesses anointed a stone with oil each day, wrapping it in wool to symbolize the stone wrapped in swaddling clothes that Rhea disguised as the baby Zeus.

Romulus Founder of Rome

The Romans identified Romulus with another god named Quirinus.

Romulus and his twin brother Remus are the mythical founders of Rome. The sons of Mars, they were raised by a she-wolf after their grandfather, King Amulius, ordered their deaths, fearing that they might grow up to depose him. The twins were thrown into the river in a basket but they did not drown and a she-wolf (wolves were sacred to Mars) suckled them. Later, the twins laid the foundations for Rome but Romulus killed Remus because he mocked the wall that Romulus made to mark out the city. When Romulus died he became a god. According to the Roman historian Livy, Romulus had gathered his troops for inspection, when a thunderous storm suddenly enveloped him in a thick cloud. When the storm subsided Romulus's chair was empty. The political leader who had been standing closest insisted the storm had raised Romulus to heaven. Thereafter he was regarded as divine and celebrated as the father of the state of Rome. Supposedly, Romulus then descended from the heavens at dawn, to announce that the gods wanted Rome to be the capital of the world and that the Romans should therefore master the art of warfare.

Senua The unknown goddess

Scholars believe that the 26 pieces of jewelery found in Senua's shrine were buried there in the third century CE to invoke the goddess's protection from a local disaster.

The Romans worshiped countless gods and goddesses, often adopted from peoples that they had conquered, which they invoked in nearly every situation they found themselves in. Archaeologists continue to unearth evidence of new cults, and scholars to research their functions within Roman society. One of the most recent goddesses to be discovered was called Senua. A shrine to Senua containing her cult statue was unearthed in a field in the English countryside as recently as 2002. Depicted as a graceful woman with her hair gathered into a bun, Senua probably is of Celtic origin but seems to have been associated by the Romans with their own goddess Minerva, probably becoming a version of that goddess in Britain. The shrine contained offerings and inscriptions thanking Senua for her past favors. She is included here as a reminder of how much information on ancient beliefs is lost to the modern reader.

Uranus Ancestor of the gods

Gaia's first child, Uranus, mated with his mother but, fearing that their offspring might take his power from him, he forced them to remain within the belly of Gaia, whose grief soon turned to hate. She made a golden sickle and persuaded their son Cronus to castrate his father. Uranus was dethroned and black blood dripped from his wound onto the earth, where it gave birth to the Furies and to nymphs. Cronus threw Uranus's genitals into the sea and Aphrodite emerged from the foam.

There is no evidence that any cult for Uranus existed in Greece. His main function, other than to personify the sky, was to sire the offspring who would become such important characters in the Greek myths.

Zeus/Jupiter Father of gods and men

The supreme god of Greek mythology and head of the Olympian pantheon, wise Zeus presided over the affairs of gods and mortals. He usurped his father Cronus in the war between the gods and the Titans and was called "the Thunderer" because he controlled thunder, lighting and rain; with just a nod of his head he could shake the very foundations of Mount Olympus, where he lived in a splendid palace built for him by the god Hephaestus. Zeus punished wrongdoing, rewarded truth and settled the arguments of the gods. Although he would sometimes intervene in the affairs of mortals, he watched events unfold on earth with a certain detachment. Zeus married Hera but partnered various mortal women, goddesses, and nymphs with whom he had countless offspring, much to his wife's anger. One of his best-known sons was the hero Herakles, or Hercules, who was Zeus' son by the mortal Alcmene.

As Jupiter, Zeus was also the supreme god of the Romans. He originated as a sky god, as he was worshiped as a rain-giver and also withholder. The Romans called him "The Best" and "The Greatest" because he controlled health, prosperity and wisdom. He presided over the city of Rome and symbolized both its might and right. Jupiter's impressive temple in Rome also had altars dedicated to his wife Hera-Juno, and his favorite daughter, Athena-Minerva. As Rome's emperors became living gods themselves, his importance decreased in Roman religion.

The games held at Olympia, a four-yearly athletic contest dedicated to Zeus, evolved into the modern Olympic Games.

Above: The Celts were originally from central Europe, but expanded into the British Isles and most of France and Spain until the Roman expansion in the fifth century BCE.

CELTIC GODS

The Celts occupied almost the whole of central Europe and parts of Gaul around 600 BCE; three hundred years later, they had spread throughout much of Western Europe, including the British Isles. Information about the Celtic gods comes from a range of sources, including the remains of ancient carvings and inscriptions, the writings of early Roman historians, the monastic records of early Irish Christians, and later, from the seventh century onwards, through literature. The epic stories were written down long after the gods had faded from belief, so sometimes the earliest evidence of Celtic mythology sits uneasily with later fictions.

The Celts regarded the land as sacred, and worshiped near springs, wells, caves, and other places that they saw as gateways to the netherworld, the world of the gods. The natural and the supernatural existed side-by-side and overlapped. The Celts had important religious festivals and feasts, which divided the year and marked the seasons. Feasts were dedicated to particular patron deities, who supposedly protected them in their daily lives.

Over time, the various gods evolved in different areas; some were specific to a particular community while others were more widespread, such as the Irish god Goibniu, the divine blacksmith, who was known to the Welsh as Gofannan. Because beliefs about the gods varied regionally, it would be impossible to assemble a comprehensive list of all the gods and many are now presumably lost to us.

When the Romans conquered the Celtic peoples, they introduced their own gods. They also gave Roman names to existing Celtic gods and imposed some of the characteristics of their own deities upon them.

Many of the gods featured here were worshiped in Ireland (which Rome did not invade) and were written about in *The Book of Conquests*, a twelfth-century text that details the mythological pre-history of Ireland and different generations of gods. It describes the gods as a series of ancient invaders who first fought for supremacy in Ireland and later colonized it.

The Fomorians were monstrous figures who, although defeated, refused to surrender and remained a threat to established order. The Fir Bolg, a later race of invaders, mythically established the ancient Irish system of monarchy but were pushed out by the Tuatha De Danaan, "the people of the goddess Dana." The Tuatha defeated the Fir Bolg by using magic and special weapons; they were adept at sorcery and skilful artisans. The Fomorians returned to fight against them: a conflict between right and wrong, order and chaos, light and dark. The defeated Fomorians were exiled so did not have the satisfaction of witnessing the ancestors to the Celts, the Milesians, rising up to defeat the Tuatha, who, according to *The Book of Conquests*, then fled Ireland. However, popular tradition held that they remained, living underground, especially in ancient burial mounds.

Right: The widely worshiped horned god Cernunnos was linked with virility and fertility. The Celts honored him in festivals, and believed he had the power to grant plentiful harvests.

Balor The evil-eyed god

A frightening adversary, Balor "of the dreadful eye" had a single eye, a glance from which could kill a whole crowd of attackers. It took four men to lift the eyelid, and Balor only opened his eye in battle. As an early sun deity who evolved into a god of death, Balor met a grim end, in fulfilment of a prophecy that his own grandson would kill him. In an attempt to circumvent this, Balor locked his daughter up in a crystal tower to prevent her from meeting suitors. Yet, through the intervention of a druidess, she was seduced and gave birth to a boy. Balor threw the baby into the sea but the sea-god Manannan fostered the infant, who grew up to become Lugh. In a conflict between light and dark, Lugh released a slingshot into Balor's eye and Balor fell dead instantly. In other versions, Lugh decapitated him or pierced him through the eye with a spear provided by the craft god Goibniu.

In Gaelic, the phrase 'evil eye' is *súil bhaloir.*

Belenus Shakespeare's Cymbeline

The Celts did not worship the sun, but they recognized it as an important power, deserving of respect. They also had a sun god, Bel, whose name meant 'shine' or "bright." He was one of the most widely worshiped Celtic gods, known as Bil to the Irish, Belunus to the Gauls, and Beli to the Welsh, and was identified by the Romans with Apollo; Julius Caesar described him as a healer, a giver of light, and a solar god. In early images, Belenus is pictured driving the sun across the sky in a horse-drawn chariot. The great fire festival of Beltane was held on May 1 to herald the summer (the victory of light over darkness) and the possibility of a successful harvest. Over time, however, Belenus's reputation diminished and by medieval times he had been downgraded to a mortal conqueror, somewhere along the way becoming identified with Bile, the god of death, who gathered up the souls of the dead and led them to the otherworld.

Shakespeare used Belenus as a source for the character Cymbeline, the English king in his play of the same name.

Brigid "The Wise Woman"

St. Bridget's feast day is February 1. Originally this was *Imbolc*, a pagan spring festival devoted to Brigid.

When the Irish converted to Christianity, this pre-Christian Irish goddess, known as "the Wise Woman" was subsumed into the character of St Bridget, the mystic midwife to the Virgin Mary. Her cult was widespread (she was Brigantia in Roman Britain) and as Brigid – the name derives from the word for "sublime" – she was an important fertility goddess, a healer who helped women in childbirth and could divine the future. She was the patroness of blacksmiths and poetry, with close links to the Celtic tradition of bardic poet-priests. Brigid was sister to Lugh and daughter of the Dagda; she had three sons and made a political marriage to Bres of the Fomorians, designed to make peace between the two warring sides. Her grief for the death of her oldest son was the first keening in Ireland. Brigid's cult was associated with holy wells and the river Brent is named after her counterpart, Brigantia.

THE CELTIC CALENDAR

The Celtic ritual calendar had four main festivals: *Samhain* (October 31 - November 1) marked the beginning of winter and the end and beginning of the Celtic year, a time when the boundaries between earth and the Otherworld were broken. It was followed by *Imbolc*, on February 1, a time of fertility rituals, associated with Brigid. The coming of summer was celebrated with *Beltane* on May 1. *Lugnasadh* on August 1 was dedicated to Lugh and linked to the harvest. In due course, all four festivals were adapted to Christian usage although the traditions of *Samhain*, in particular, have survived as Hallow'een and All Souls' Day.

Ceridwen Mother of Taliesin

The Halloween image of an old crone stirring a bubbling cauldron of witch's brew has its origins in the story of Ceridwen.

The Welsh crone and prophetic mother goddess Ceridwen brewed divine knowledge and inspiration in the cauldron of the underworld. She boiled up wisdom for her revoltingly ugly son Afagddu, to compensate for his poor looks. She asked a young boy, Gwion, to guard the cauldron, which needed to be boiled for a year and a day; as he did so three drops of the brew spilt on his finger, which he then unwittingly sucked. As a result, he understood all the world's secrets, could hear everything in it and had the foresight to know that Ceridwen, in a rage at the wisdom he had acquired, would pursue him. He fled, she chased, both repeatedly changed shapes, until Ceridwen in the form of a hen swallowed Gwion as an ear of wheat. This made her pregnant and nine months later she gave birth to the great and beautiful Welsh poet Taliesin, who was a reincarnation of Gwion in this mythical version of the genuine fifth-century poet's origins. The word *taliesin* means beautiful forehead. It is a story of renewal, rebirth, and initiation and led to Ceridwen being revered as a goddess of fertility.

Cernunnos "The horned one"

Revered from Romania to Ireland but chiefly associated with the Gaelic Celts, the horned god Cernunnos was the lord of animals, and could turn into one in an instant. A Pan-like figure, he was associated with the tasks of hunting and culling and his responsibilities extended to agriculture, since he also ensured a generous harvest and the prosperity that came with it. He was pictured with a body of a man and the head of a stag, often seated with his legs crossed as hunters do. Well into Christian times, people would wear stag costume at festivals to honour him. His horns symbolized virility and fertility, and he was often accompanied by horned snakes – snakes slough their skin, signifying rejuvenation. He is recalled in UK place-names such as Cerne Abbas and he was probably analogous to the legendary "Herne the Hunter." The Christian church condemned Cernunnos as an evil figure of sexual excess; in the Middle Ages, he was equated with the devil.

A carving in Northumberland, England depicting Cernunnos is over 50 metres (160 ft) high; much smaller ancient carvings of Cernunnos are very common, often with holes in the head for his antlers.

Dagda, The "Father of all"

The Irish Celts called the Dagda "Great Father," "Good God," and "the Mighty One." He was the head of the Tuatha De Danann (a dynasty of gods), whom he led into Ireland, and was the father of Brigid and the patron of druids. This Irish earth god dressed in a simple rural tunic so short one could see his bare buttocks, and dragged a huge club behind him on a set of wheels. His ludicrous appearance and his club suggest echoes of an earlier fertility god. The Dagda was an impressive fighter and used one end of the club to kill his enemies; with the other he could bring the dead back to life. He had an inexhaustible cauldron, so he could always feed his people (although oath breakers and cowards went hungry), two pigs, one always roasting, the other always growing, and a magic harp made from a living oak, which summoned the seasons, playing music of joy, sadness, and dreaming. Adept at magic, the Dagda was both knowledgeable and wise. He had a rapacious sexual appetite and his annual mating with his consort, the Morrigan, ensured the security of his people.

The large naked figure cut into the chalk ground at Cerne Abbas in Dorset, England is similar to the Dagda but there is no other evidence that he was worshiped outside Ireland.

Danu "The waters of heaven"

Don, a version of Danu, is incorporated in the names of many European rivers: Don, Dnieper, Donwy.

The ancestor of all the gods, Danu (also known as Don and Danann) was an earth mother and fertility figure dating from a remote period in pre-Christian Ireland. In some accounts, she is said to be the wife of the Dagda, others hold that she was his mother or his daughter. As the mother of the gods, Danu was a life-giving force; she had a protective interest in rivers, streams, and wells, and gave prosperity to her people. Although there were malicious aspects to her character, her offspring, the Tuatha De Danaan ("the tribes of the goddess Danu"), represented the forces of good and battled the dark, evil Fomorians, whom they replaced as mythical inhabitants of Ireland. When the Tuatha De Danann were defeated by the Milesians (Sons of Mil) they retreated underground and, in popular tradition, turned into *sidhe* or "fairies." Danu was demonized by Christian missionaries as a crone who captured and ate children.

Dian Cecht's name was associated with medicine until the eighth century CE at least.

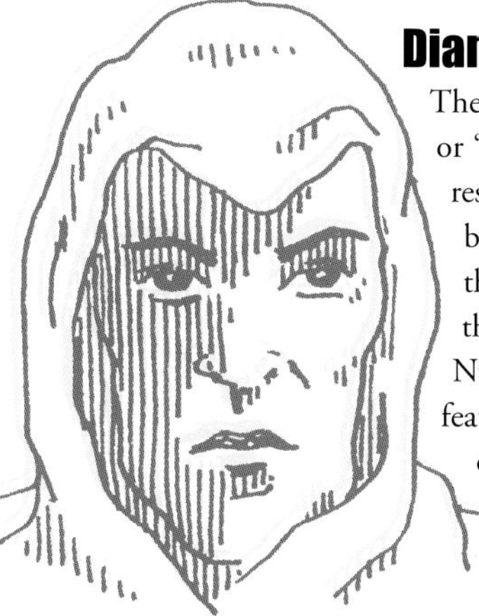

Dian Cecht The physician

The Irish god of medicine, known as "the Leech" or "the Sage of Leechcraft," Dian Cecht could restore the mortally wounded to health and bring dead warriors back to life by immersing them in a healing well and chanting spells over them. He was also a skilled smith and replaced Nuadu's missing hand with a silver replica that featured moving fingers. His son, Miacht, outshone Dian Cecht's work by supplying Nuadu with a real hand. In jealousy, Dian Cecht killed him. Concerned that his daughter Airmid, known as the "she-leech," might also outdo him, he prevented her from cataloging all the magical healing herbs that grew from her brother's grave by constantly disturbing her. He was the paternal grandfather of Lugh, through his son Cian, born of a liaison with Danu.

Donn "The Dark One"

A silent, isolated and aloof figure, Donn was the Irish god of the dead. Known as "the Dark One," he lived in the "House of Donn," which served as a meeting place for the souls of the dead to gather for their journey to the otherworld. When Donn drowned in a bay in Kerry, it was his dying wish that all his descendants should assemble there. The Irish regarded him as an ancestor figure and believed that to see him was a sign of imminent doom. Presumably because of the manner of his own death, they associated him with shipwrecks, storms and other seafaring misfortunes. The Romans linked him with Dis Pater, mentioned by Julius Caesar, who was the Roman underworld god and ancestor of all men.

Christian legends demonized Donn, equating the "House of Donn" with purgatory, a place of after-death suffering.

THE OTHER WORLD

The Celtic druids, or priests, taught that the soul is immortal and that it lived after death in the Otherworld. When one died there, one returned to this world. Death in this world therefore involved a rebirth in the Otherworld, and a birth in this world signalled that there had been a death in the Otherworld. Some Otherworldly locations were paradise (like Avalon) whereas others (like the House of Donn) were places of shadow and misery. The Otherworld was the abode of the gods and had close links to this world. Mortals needed permission to enter it (male heroes were often invited in by beautiful and seductive Otherworldly maidens) though the gods needed no invitation to enter the mortal world. The Celts believed that various features of the landscape were the entrances to the Otherworld, and often built shrines to the gods at springs, caves, wells, and rivers, all of which they considered to be gateways to a divine realm.

Epona "The 'Divine Horse"

In Ireland, the horse goddess Epona was the bringer of dreams. Nightmares were associated with her, although she brought good dreams too. Yet Epona stems from Gaul, where she was particularly important in rural areas, being revered as "Divine Horse." In early carvings, she is shown on horseback, usually riding sidesaddle, accompanied by foals and dogs. She often carried a bowl of fruit and ears of ripe corn, which suggests she was closely linked to fertility and abundance. She sometimes appeared as a beautiful naked nymph at waters that were believed to have healing properties. She was the only Celtic god worshiped in Rome, where she was important enough to be given an official feast day on December 18. The Roman cavalry had statues of Epona in their stables to protect their horses, and even their donkeys and mules; the troops spread the cult of Epona wherever they went, and she was worshiped throughout Gaul.

Epona's name is linked with *epos*, a Celtic word for horse.

Goibniu Brewer of the gods

The Celtic concept of Goibnu was influenced by the Roman idea of Vulcan (derived from the Greek god Hephaestus), a divine craftsman who was also a brewer.

Goibnu brewed the beer of the gods in a massive bronze cauldron, and hosted a feast at which he encouraged them to drink a magical concoction. Instead of intoxicating them, the beer protected them from ageing, disease, and death.

Goibnu was also a metal smith and armourer of the gods, making and repairing weapons with three strikes of his hammer. His weapons always struck true and his spears were especially deadly. Goibnu was one of a triad of craft gods, with Credne (brass and metal worker) and Luchta (wheelwright), and had healing powers.

Lugh "The Long-armed"

The word leprechaun (fairy-like guardian of treasure) comes from Lugh's name. And the word London (Latin: Londinium) may derive from Lugdunum.

Lugh is probably the best known of the Tuatha De Danann gods, although he was descended both from the Dagda, head of the Tuatha, and from Balor, leader of the Formorians. He was also the father of the great hero Cuchulainn. Known as the lord of skills because he excelled at just about everything, Lugh once introduced himself as a poet, historian, physician, cupbearer, craftsman, and builder. He was also called Lugh Lamfhada (Lugh of the long arm) because of the huge distance he could throw a weapon. He is best remembered as the Irish sun god (the equivalent of Lleu in Wales, and Lugos in Gaul), an athletic and majestically handsome warrior who entered Celtic mythology fairly late and was identified by the Romans with Mercury. As a god of light, Lugh's summer festival was *Lugnasadh*, which is the Irish name for the month of August. When Nuadu was injured, Lugh replaced him as king of the Tuatha De Danaan. During the war between the Tuatha and the Fomorians, Lugh killed his one-eyed giant grandfather Balor with a slingshot delivered at such close range and with such force that it pushed Balor's eye to the back of his head so that it glared out on Balor's own troops behind the giant, and slaughtered 27 of them. Lugh was credited with inventing the board game *Fidchell* – an early Irish game that is now lost to us outside descriptions in literature, but which may have had similarities to chess. He also introduced the sport of racing to Ireland.

Mabon An Arthurian hero

A Celtic Welsh god of youth, the young warrior Mabon became the subject of one of the legendary king Arthur's heroic adventures when Arthur, equipped with animal wisdom, rescued him from a magical prison. Mabon was associated with youthful pursuits, especially hunting, and in one myth he killed the magical boar, Twrch Trwth. The Roman version of Mabon, Maponus, was an archetypal "Young Man," often depicted, like Apollo, standing tall and naked and holding a lyre.

Although his name came to mean "Great Son," it originated in an Irish word meaning, simply, boy.

Manannan "The son of the Sea"

The best known and most important sea god, Manannan rode the waves as though they were horses and had a boat called "the wave-sweeper" that sailed itself without oars or a sail. In another account, he had a chariot pulled by a white horse, and crossed the water as though it were land. Manannan was lord of the sea but he also had important responsibilities in the Otherworld, where he ruled the Land of Promise, a mythical paradise closely related to the Avalon of King Arthur. Manannan carried his possessions in a skin bag; they included the letters of the alphabet and a cloak that he could wave between people to ensure that they never met. It changed destinies and was as varied in color as the sea. He also owned a sword called "the Answerer," which could pierce any armour, and a breastplate that offered absolute protection. He had animals that could be slaughtered and eaten, coming to life again for the next feast. Manannan was also a legendary shape-shifter, adept at magic, and wandered Ireland in many different forms, often playing harmless tricks on mortals. *Inis Manann*, the Isle of Man, was named after him and he was its first king.

Manannan was patron god of the Isle of Man; the island is named after him.

Math Magician and king

In the legends about Saint Patrick, Math became a pagan prophetic figure who spurned Christianity.

The great magician and king Math was the brother of the mother goddess Danu (or Don) and was the god of increasing prosperity. His own continued existence depended on his feet being held in the lap of a virgin, although he was able to leave this resting place to do battle. His brother Gwdion encouraged Math to go to war, leaving the young woman to succumb to the advances of Math's nephew, who loved her. Math returned furious and turned his nephews into animals in a display of brutal revenge and impressive magic. In successive years, they became stags, boars, and wolves, fated to mate and have offspring with one another. It is likely that the virgin foot-holder represents a sacred bond between the king and the land.

Morrigan, The "Washer at the Ford"

The word *morrigan* literally meant "mare queen," and came from the words for "nightmare" and for "queen." The Morrigan is also thought to be the forerunner of Morgan la Fée in the Arthurian cycle.

The Morrigan, the great "Phantom Queen" of the Irish Celts and the goddess of death, embodied all the furies of war. She specialized in slaughter and used her terrifying appearance to destroy her enemies. As a shape-shifter, she could combine an alluring and deadly sexuality with her awesome powers of prophecy and spell casting. When the Celtic hero Cuchulainn refused to make love to her, the wound she gave him sealed his fate. Cuchulainn met her again in the guise of the Washer at the Ford, a maiden who told him she was washing the blood off his armour: a chilling foretelling of his death. After the hero was killed, the Morrigan sat on his shoulder in the form of a crow, watching a badger drink his blood. She often scavenged the corpses herself or offered the blood of her victims as a gift to their enemies. When the Morrigan left her cave-home to appear as an old hag it signaled the imminent demise of a ruler.

The Morrigan was one of three war goddesses – the others being Nemain (Raven) and Badb (Frenzy) – who may have been aspects of her triune personality. Yet she had positive qualities too. In some traditions, she was regarded as a fertility goddess. As the sexual partner and wife of the Dagda, she mated with him at the start of the year to ensure prosperity.

Nuadu "Silver-handed"

In Irish mythology, rulers were not allowed to have physical imperfections, so when king Nuadu lost his right hand in battle he also lost his kingship. After seven years, the medicine god Dian Cecht made him a magical silver replacement, and he became known as Nuadu Argatlam, Nuadu Silver Hand. He then regained his regal authority as an early king of the Tuatha De Danann and their military leader. He had a deadly sword from which no one could escape or survive, and played a leading role in ancient stories that addressed issues of sacred kingship and the protection of land.

A mysterious and prophetic figure with healing powers who probably originated as a tribal war god, Nuadu died in battle. According to Irish legend he was buried in Donegal, Ireland, at the stone fort of Ailech, west of Derry, which was one of the fortresses of the Tuatha De Danann.

His cult centered at Lydney, Gloucestershire in England, later being taken over by the Romans, who renamed him Nodons and associated him with healing and water.

Oghma The god of eloquence

As the god of eloquence, Oghma wielded considerable power: he was easily able to influence and impose his will on others through his "honey-mouthed" speaking. This orator-warrior reputedly invented the first Irish alphabet so that he could encode knowledge, but was best known for the fine golden chains that ran from the tip of his tongue to the ears of his mortal listeners, a symbolic representation of the spell-binding power of his rhetoric. The Celts believed that eloquence was of greater power than physical strength. Writing and eloquence were high-status skills, often linked to sacred rituals, and Oghma taught his skills to the priests, so he was probably an important and revered deity for the Celts. In addition to his bardic role, he accompanied the dead on their journey to the Otherworld. The Irish Celts depicted him as an old man with a bald head and wrinkled dark sunburned skin, from which came his nickname, "sunny faced," a term that may also allude to his bright intelligence.

The ogham alphabet was Ireland's earliest form of writing and examples still survive, cut in stone or wood.

THE OGHAM ALPHABET

Ogham (from *ogmos*, Greek for "furrow" or "straight line") was an ancient Celtic form of writing on monuments, traditionally invented by Oghma. One to five lines (consonants) and points (vowels) are made in relation to a guideline. The ogham alphabet has only 20 letters. Inscriptions have been found in Ireland, on the Isle of Man, in Wales and in Cornwall.

NORSE GODS

Between the third to sixth centuries CE, Germanic tribes settled in Britain and Scandinavia. They were known as the Northmen or Norse. In Britain they were exposed to the Christianity and coverted, but in Scandinavia, from the eighth to the eleventh centuries, when they finally abandoned paganism, the Norse, better-known as the Vikings, worshiped a large pantheon of gods, each with their own mythology.

Above: The Norse – or Vikings – were from the Scandinavia and voyaged extensively across the North Atlantic and as far south as the Black Sea.

The Vikings initially had two different sets of gods: the old Germanic ones that were predominantly fertility figures, and their own new ones that – like the people who worshiped them – were chiefly warrior gods.

Myths emerged that explained the two different sets of gods and merged them into the same system. Mythically, the Vanir (the old Germanic gods) fought with the Aesir (the Norse gods) for supremacy. The Norse believed that the Aesir lived in the divine city of Asgard, which they situated in heaven, and that, although the walls of Asgard were destroyed in the war, the city was not. Neither side emerged victorious and peace was declared. To ensure that peace prevailed, both sets of gods exchanged hostages so, for example, the old Vanir god Freyr joined the Norse gods in Asgard.

The Vikings believed that the main gods had their own halls, or palaces, in which they lived. Valhalla, Odin's home, was the most important. It was heaven to Viking warriors who had been slain in battle, who were escorted from the battlefield by the Valkyries, divine maidens who continued to attend them in heaven. There the warriors drank, feasted, and jousted their days away.

Much of our information about Norse beliefs comes from an unreliable source, the Christian scholar Snorri Sturluson, who wrote the Norse myths down in the *Prose Edda* in the twelfth or thirteenth century. He had his own Christian beliefs and he downplayed the power ascribed to the Norse gods to ensure that he did not upset his Christian readers, reducing gods like Thor and Odin to war-chiefs rather than divinities.

In the Norse pantheon Odin, Thor, and Freyr were probably the most important figures. Odin was a relative latecomer and archaeological excavations suggest that worship of Thor and Freyr was originally more extensive than that of Odin.

Right: Tyr originated as a Germanic sky god and emerged in the Norse pantheon as a warrior figure renowned for his bravery. He was a popular figure in Norse poetry, in which he was frequently celebrated for his wisdom.

For many peoples, gods are by definition immortal. This was not true for the Norse, who believed that it was the gods' access to magical apples that kept them young. They also envisaged that many of the gods would die. Balder himself died through the mischief-making of the god Loki, but many more would fall at the Doom of the Gods, Ragnarök, when the forces of good (aided by Odin's slain warriors) and evil would meet and fight. Doomsday itself would lead to a new world order of prosperity and ease.

Aegir God of the sea

The sea ruled the lives of fishermen and Aegir ruled the sea. He personified the ocean and its power, both good and bad. He called up storms to wreck ships and he and his wife, Ran, worked a net and trident to pull the drowning down to his underwater kingdom where he feasted the other gods. It was lit by the sparkling of gold and sailors often traveled with a piece of gold so that Aegir would look upon them kindly if they drowned.

His nine daughters had names like "howler" and "grasper" and were doubtless personifications of waves. Ran, his wife, was herself a personification of water, as waves flowed from her mouth. Aegir is depicted as an old man with grasping fingers and a long white beard, suggesting the trail behind a boat, perhaps, or a trail of breaking waves.

The Northern Europeans called a stormy sea "Aegir's cauldron," in reference to a myth in which Odin made Aegir brew ale for the gods.

Balder God of light

The god of wisdom and light, Balder, "the best of all the gods," was merciful and softly spoken. He lived in the magnificent hall of Breidablik, which reflected the glory of his character – nothing impure was allowed to enter. The son of Odin and Frigg, he was adept at magic. When Balder foresaw his own death in bad dreams, Frigg asked every living thing to swear not to harm him (a promise all were happy to make). However she overlooked the mistletoe, and when Loki persuaded Balder's blind brother, Hoder, to throw a branch of mistletoe at Balder in a game it pierced and killed him instantly. Balder joined Hel in the underworld and an otherwise unanimous decision by the gods that he be released was frustrated because Loki (hiding in a mountain cave disguised as an old woman) dissented. After Ragnarök and the destruction of the gods, Balder emerged to rule over the green and pure land of a new world order. Hardly surprisingly, Christians identified him with Christ.

Balder was a popular subject in nineteenth century paintings and sculptures, which often depict him standing in a Christ-like pose, palms extended.

Bragi The patron of poets

Bragi was subsumed into a real-life ninth-century poet, Bragi Boddason, whom later poets established in the pantheon of gods.

Son of Odin, the white-bearded Bragi was the god of music and poetry. He married Idun, keeper of the apples of immortality. As the bard of the gods, he played a magical golden harp and declaimed their great deeds in a wonderful voice that could move heaven and earth. He personified wisdom and eloquence and had runes carved on his tongue. Mortals swore oaths in his name. In one myth, Loki accuses Bragi of killing Idun's brother; in another, he suggests that Bragi was a coward. After Balder's death, Bragi confronted Loki, threatening to cut off his head. Loki fled, foretelling Ragnarök, the doom of the gods. Bragi therefore played a key role in Loki's transformation from a mischievous figure to one of evil.

Forseti The peace broker

Forsetlund, near the Oslo fjord, was named after this obscure god, suggesting that he was worshiped there.

The son of Balder, Forseti was the god of law and justice. He acted as the divine judge for both gods and mortals from his seat of judgment in the heaven of Asgard. His vast hall, Glitnir, had a silver roof that was supported by pillars of red gold. All who went to Forseti with a dispute found themselves provided with the resolution of their argument. The gods swore oaths in his name and did not dare to renege on them.

Freya The lady of love

Scholars often link Freya with the Egyptian goddess Isis, partly because Freya wore a dress of feathers that could turn her into a falcon, and Isis was a winged goddess.

The sensual goddess of sex who was also connected with war and death, Freya was Freyr's twin sister and one of the Vanir, going with him and Njord to Asgard as hostages after the war with the Aesir. When deserted by her unfaithful husband Odur she wept tears that turned to amber and gold,

her desperate love leading to a desperate grief and to bouts of promiscuity. Freya was patroness of childbirth and of lovers, often being invoked in affairs of the heart. She journeyed in a chariot drawn by two cats and was leader of the Valkyries, maidens who flew over battlefields and collected slain warriors, leading half to Valhalla (the hall of the slain) and the rest to Freya's hall, Folkvang, where they were reunited with the women they loved. In later myths, she is merged with Idun.

Freyr The lord of the harvest

A good-looking and powerful fertility god, Freyr was lord of sun and rain and controlled the harvest. He was a brave warrior, one of the Vanir race of gods, and went to Asgard with his father, Njord, and his sister, Freya, as hostages to end the conflict with the Aesir, becoming known as "Lord of the Aesir." Even for a god, Freyr was extremely well equipped. He journeyed in a magical ship called Skidbladnir that never had to search for wind and which, although it was big enough to hold all the Aesir and their weapons, folded up to fit in a pouch. He also had a chariot pulled by a golden boar, a magic sword that leaped into battle as soon as it was drawn, and a fearless horse. Nevertheless, he was destined to die at the doom of the gods, being unable to fight with his magic sword, which he lost in his attempts to woo Field (Gerd).

Freyr was widely worshiped in fertility cults, and statues show him with a large penis, suggesting he was seen as a very virile god.

Frigg Queen of the gods

Frigg originated as an earth mother and became the goddess of conjugal love and of marriage. She offered hope to the grieving, ensured fertility and was patron of the childless. A tall and beautiful woman, she was Odin's wife and goddess of the sky, in whose colors she dressed herself, from the white of the clouds to the darkness of the night and the brooding shades of stormy weather. Although Frigg could descry the future, she never revealed what she saw. She was often pictured at a spinning wheel from which she spun golden clouds. In her welcoming hall of Fensalir she reunited dead husbands with their wives.

The Germanic version of Frigg is Frija, who gave her name to Friday.

Heimdall The divine watchman

Snorri Sturluson refers to a poem called the *Heimdallargaldr*, now lost, which may have recounted Heimdall's character and deeds in full.

Born of the sea, or supposedly the son of all nine daughters of the sea god Aegir, Heimdall acted as the watchman of the gods. He guarded Bifrost, the bridge that linked the earth with the heaven of Asgard, to ensure that the giants would not cross it. This shape-shifting god lived in Himinbjorg ("The Cliffs of Heaven"), on the Asgard side of the bridge, from which he could keep watch, armed with a sword and his impressive powers. His hearing was so acute he could hear the grass grow and the wool grow on sheep. Needing less sleep than a bird (in other words, none) Heimdall could see at night, and across huge distances. His horn Gjallar signalled the doom of the gods with a sound so loud it could be heard throughout all the worlds; it alerted the gods to Ragnarök, when he and his great enemy Loki were destined to kill each other and the world of the gods would be reduced to ashes.

Hel Queen of the underworld

The English word "hell" clearly derives from the Scandinavian *hel*. Christian writers turned Hel into an increasingly monstrous figure and her realm of the dead was seen as a place of punishment for the wicked.

The daughter of Loki, Hel ruled the land of the dead, the dark underworld realm of Helheim, a kingdom enclosed by a high wall and gates guarded by a dog, which in some accounts had to be reached by journeying across water. Helheim became home to mortals who had died from old age or disease, rather than meeting a heroic end in battle. The *Eddas* suggest that there were different realms of Helheim, and that the ninth – Niflhel, or Misty Hell – was the last and most remote, a place of punishment for adulterers, murderers, and other evildoers. These beings would rise up to fight against the gods at Ragnarök. Hel herself made a frightening figure, a grim-looking hag. She is usually described as half-alive, half-dead, probably because parts of her body were decomposed, whilst others were living flesh. At Ragnarök, she and her army of ghostly warriors fought against the gods but ultimately flames destroyed her underworld realm.

Hoder The blind god

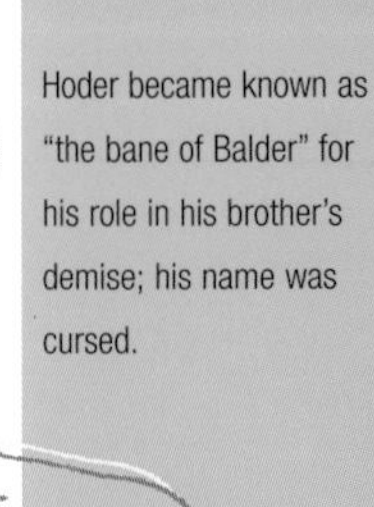

The god of darkness, Hoder, blind from birth, was the son of Odin and twin brother to Balder, the god of light. Loki tricked him into throwing mistletoe at Balder, which killed him, although the other gods blamed Loki more than Hoder, the implication being that Hoder imagined the mistletoe to be harmless. However a very different and earlier poetic account tells how Hoder and Balder competed with each other, fighting several battles for love of the maiden Nana. In some versions of this tradition, Hoder marries her but still goes on to kill the heartbroken Balder; in others, it is Balder who marries Nana. Both versions supply a motive for fratricide; Loki may have chosen well in using Hoder as an unwitting accomplice – perhaps the god of darkness was blind to his own actions. Certainly, this immensely strong god dispensed good luck or bad fortune in an ad hoc manner. Hoder was reborn into the new world that arose after Ragnarök.

Hoder became known as "the bane of Balder" for his role in his brother's demise; his name was cursed.

The mythical great world tree, Yggdrasil, was firmly rooted in Hel's underworld abode. Its branches spread out across the entire world of mortals, known to the Norse as Midgard, and up into the heavenly Asgard, where the gods held their daily council beside its branches. The three Norns, deciders of destiny similar to the Greek Fates, sustained the tree and its rotting trunk by sprinkling it with the life-sustaining waters from the Well of Fate that lay at its base. An eagle perched at the top of the tree, causing the earth's winds to blow by the flapping its wings. At the foot of the tree a serpent named Nidhoggr ate at the roots. The world tree was the pillar of the universe, and serves as a symbol of the interconnectedness of life.

Idun Keeper of the apples

The goddess of eternal youth and of spring, Idun was the wife of Bragi, the god of poetry. She was associated with fertility and youth, and guarded the apples of youth for the gods, carrying them in a box that was never empty. The apples ensured immortality, without them the gods would wither into old age. Loki once tricked her into abandoning the apples by persuading her that there were better ones nearby. In her absence, he stole the rejuvenating apples and the gods grew old and weak until Odin forced Loki to make restitution and the youth of the gods was restored.

The ancient Greek poets were the first to establish the link between apples and gods. Hera's magic apples grew in her orchard on Mount Atlas where they were guarded by the Hesperides. Scholars debate the extent to which the Greek myth provides a source for Idun's responsibilities.

Loki The maker of mischief

In Norse mythology, Loki's character evolved from being merely mischievous to the embodiment of evil.

The son of a giant and probably originally a fire god, Loki developed from a restless, mischievous figure to one responsible for the death of Balder and leader of the forces of evil at Ragnarök. A clever and witty trickster, although constantly being forced by the other gods to make amends, Loki was a shape-shifting master thief with impressive magical powers. He was also the "Father of Lies" and father of the dark goddess Hel. The late Vikings linked him to the devil, and according to the Prose Edda of Snorri Sturluson (a thirteenth-century writer who compiled a detailed record of pre-Christian tradition) he was punished by the other gods for his role in the death of Balder. They bound him with the entrails of a wolf and poured the poison from a serpent on him. His struggles to escape caused an earthquake and led to Ragnarök, the end of the world of the gods.

Njord The spirit of the waters

Around 100 CE, the Roman writer Tacitus wrote a detailed account of an earth goddess called Nerthus, worshiped by the Germanic tribes. Linguistically, the name Njord relates to Nerthus, so Njord may be based on her.

The god of sea, wind, and fire, Njord could calm stormy waters, still wind and control fire. Seafarers linked him closely with fishing and sea travel, and worshiped Njord in the hope that he would bring them safety and prosperity. One of the Vanir gods, Njord led a successful attack on Asgard, and also defeated Loki's attempt to usurp him as leader. After the war between the Aesir and the Vanir, he became a hostage to the Aesir gods. He fathered Freyr and Freya by his sister, and his home consisted of an enclosure of ships. He was briefly married to Winter, Skadi, but she wanted to live in the mountains and he beside the sea. They tried to compromise by dividing their time between both homes, but this proved unsuccessful: Njord hated hearing the howling of wolves and Skadi hated hearing the crying of seagulls, so they separated.

Odin Father of the gods

Odin's many titles make his importance clear: from "giver of victory" and "lord of the slain" to "all-father" and "all-knowing," his powers were remarkable. He ruled heaven and earth, sky, wind and war and was lord of life and death, greeting warriors killed in battle at Valhalla, the hall of the slain. Odin particularly prized knowledge: he is often depicted as an old one-eyed man, having exchanged the other eye for wisdom and the knowledge of death, magic, runes, and the occult. He gave weapons to those he trusted; in return, they swore fealty to him until death and beyond. Odin could be fickle and treacherous, however; he instigated wars and gave strength to mortals in battle but he might also withdraw his support unexpectedly. Sometimes his shifting decisions seem to have been quite arbitrary, as if his main concern was to ensure that blood was spilled. Certainly, the northern Europeans conducted human sacrifices in his honor. Two ravens, Mind and Memory, sat on his shoulder, flying off every morning to gather all the world's news, which they then whispered in his ear as he surveyed his dominions from a magnificent hall in the heaven of Asgard.

Odin traveled through the air on his eight-footed steed, Sleipnir. He carried a magical ash-wood spear, which never missed its mark. His ring, Draupnir, symbolized his strength and power but it could not save him at the final battle, Ragnarök, the doom of the gods, where he was fated to die.

Odin is derived from the Old Icelandic word odur, or furious. He is Wotan in Germanic and Woden in Anglo-Saxon. Wednesday is Woden's day.

Sif Goddess of thunder

The beautiful wife of Thor and the goddess of thunder, the importance of Sif within the pantheon of gods is unclear. She is best known for being the victim of a cruel joke when Loki cut off her long golden hair as she slept. To appease Thor, who was angry at Sif's distress, Loki then caused a golden wig to be made for her. Its fine strands of gold magically grew on her head, stirring at the lightest breath. In another, much later, narrative Loki boasts that he had slept with Sif, an action that makes the gods realize that Loki will soon be their enemy.

Poetic descriptions liken Sif's hair to ripe corn, an image of harvest and abundance that suggests she had a role as a fertility goddess.

Thor God of thunder

Thursday is named after Thor.

The mighty warrior Thor (Germanic: Donar) was second only to Odin. A heroic figure with a great appetite, he was the strongest of all the gods, Thunder was caused when he hurled his hammer; when he struck it against stone lightning flew like sparks. In differing accounts, he made thunder by blowing into his beard or as he crossed the heavens walking alongside his chariot, whose iron wheels made thunder as they turned. Thor's hammer Mjollnir always returned to his hand; it could shatter the skulls of giants and raise the dead from their graves. Not simply a destroyer, Thor brought good fortune to brides and newborn babies in an extension of his original role as a fertility god, although the sagas lightly mock his boisterousness and lack of intelligence.

Tyr God of battles

Tyr was known to the Anglo-Saxons as Tiw. The word Tuesday stems from Tiw's day.

Tyr (Germanic: Tiwaz or Tiw) was originally an important sky god. His father, Odin, overtook him as the supreme deity and many of Odin's characteristics resemble those originally ascribed to Tyr. He was the wise and bold god of battles and of war, concerned with justice. Tyr died at the doom of the gods when he and the watchdog Garm (who guarded the gates of Hel) killed each other. The manner of his death echoes a myth in which he lost his hand to Loki's wolf-son Fenrir. Tyr usually fed Fenrir and placed his hand in the wolf's mouth as a gesture of trust, to deceive Fenrir into allowing the gods to place a chain around its neck. Fenrir was not deceived, however, and bit Tyr's hand off.

Uller The patron of skiers

The son of Sif and stepson of Thor, Uller was the god of death and winter. He was probably of Germanic origin, although later myths describe him as coming from the Vanir lineage of gods and crossing to the family of Aesir to live at Asgard after their early warring. He excelled in skiing and archery and lived in Ydalir, the dale of yews, an appropriate home as archery bows were made from yew wood. He was also an accomplished warrior, particularly strong at duelling, and was called "the god of the shield;" shields were known as "the ships of Uller," implying that he used his shield as a boat. The term also suggests that he was a protector figure, who shielded his followers from their enemies. Another tradition has him crossing the sea on a magic bone, which might refer to ice skates or skis. It is clear that as the god of winter he traveled particularly well on snow. He married Skadi, who personified winter.

Uller's name is closely related to a Germanic word for majesty, which may support the theory that he was an old Germanic sky god.

Vali The avenger

We know relatively little about Odin's son Vali, the god of archery, of marksmanship and of battles. From the information that survives, it is clear that his whole reason for being was to avenge the death of Balder by killing the blind Hoder: he grew from a baby to an adult in one day. A courageous fighter, he survived the doom of the gods to become one of the rulers of the new era that followed. He is often wrongly confused with another Vali, the son of Loki, who in some accounts played an unwilling role in the demise of Loki

Vali and St Valentine were both archers, but whether they share a common root is less clear.

Vidar The god of silence

The son of Odin and a giantess, Vidar was a silent, solitary and somewhat mysterious figure who lived in the hall of Vidi, described as being in a calm clearing surrounded by tall grasses and trees. This suggests that he may have been a forest god associated with fertility and new life. He was physically one of the strongest gods and highly dependable, being destined at Ragnarök to avenge the death of Odin, who was swallowed by the wolf Fenrir. Placing one foot on the beast's jaw, Vidar tore the wolf apart, surviving to become one of the leaders of the new world.

The poet Snorri Sturluson encouraged his listeners to throw away their old shoes, as Vidar's shoes were made from the scraps of leather that came from them. Since Vidar's shoe would eventually stand on the head of Fenrir, this was a way for ordinary mortals to assist the divine forces of good.

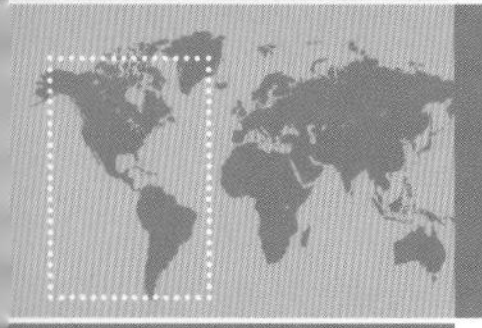

americas

Right: Cinteotl – the Aztec god of maize: a sacred, precious crop and considered the basis of all life.

Below: The Ancient Mayan ruins at Palenque in Mexico set in the picturesque foothills of the Tumbala mountains.

The Americas were settled by waves of immigrants who came from Asia across a land bridge that is now the Bering Straits between eastern Siberia and Alaska. The date at which the first migrations began is still hotly debated. Less controversial is the date at which the migrations ended, when the rise of sea levels at the end of the last Ice Age (c. 8000 BCE) submerged the overland route. The continent was isolated until its rediscovery by Christopher Columbus in 1492. During the long millennia of isolation, the native peoples of North, Central and South America evolved many different civilizations, from the sophisticated urban cultures of Mexico and the Andes to the nomadic hunter-gathering peoples of the Great Plains of North America.

Past writers, who refused to believe that Native American peoples were capable of developing sophisticated urban civilizations, attempted to prove that there were links between the Old and New Worlds. The nineteenth-century European explorers of the Maya world, for example, when they re-discovered the Maya pyramids, calendar, and hieroglyphic writing, suggested a direct link with Egypt. However, these resemblances are now seen as purely coincidental, and their separation in time as well as distance makes such a link impossible. Scholars have not ruled out that America was visited by the Vikings or Irish monks between the eighth and eleventh centuries, but these contacts left little or no trace and did not influence the development of native cultures in Central and South America.

Above: Crowds watch from the ancient walls of Sacsayhuaman, Peru as women pray during the Inca winter solstice festival, Inti Raymi.

The Aztecs of Mexico and the Inca of Peru, who were destroyed by the Spanish conquistadors in the sixteenth century, were the last in a long succession of Native American cultures that had dominated Central and South America for two and a half millennia, and included the mysterious Olmecs, the Maya, the Nazca and the Chavin peoples. These sophisticated urban civilizations shared many common features: elaborate polytheistic pantheons of gods and goddesses, the institution of divine kingship and a belief in the power of human sacrifice to preserve the world and the gods from destruction.

North America, although it was reached first by groups of migrants crossing over the land bridge from Asia millennia before the end of the last Ice Age (c. 8,000 BCE), did not witness the emergence of complex urban civilizations or of great empires.. Many of the Native North American peoples were hunter-gatherers whose religious beliefs were based on a harmonious relationship with nature, the spirits, and the gods. The nomadic cultures of the Great Plains, which we associate with Native American life, only emerged in the seventeenth century, after the horse, which had become extinct in the Americas in prehistoric times, had been re-introduced by European settlers.

AZTEC AND MAYA GODS

The area that is now Mexico, Guatemala, and Belize witnessed the emergence of many great civilizations. The Olmecs thrived from around 1500–400 BCE. Later came the Zapotecs and the Maya, then the Toltecs around 900 CE. When natural disasters and war led to the collapse of the Toltec empire, several peoples, including the Aztecs, rose to power in Mexico. By the mid-thirteenth century, their empire had absorbed the old gods of the Toltecs and other conquered peoples alongside their own.

Above: The Aztec territory was mostly in the west of modern Mexico while the Maya were more in the east and parts of Belize and Guatemala.

The Classic period of Maya civilization lasted from 250–900 CE, when the Maya built large cities in the jungle interior of the Petén region of Guatemala. Once thought to be ceremonial centers inhabited by priests, the Maya cities are now known to have been populated by tens of thousands, who lived in simple wood-and-thatch houses, around the brightly painted stone and stucco palaces and pyramids of their divine kings. Although they did not invent the Mesoamerican calendar, astronomy, and picture writing, the Maya developed these to their highest forms in the Americas. For all their sophistication, the Maya's religious beliefs still strike us as strange and violent. Like many other Native American peoples, they believed in the efficacy and necessity of blood sacrifice to ensure that the gods and cosmos endured. This blood sacrifice was obtained from self-inflicted blood-letting rituals and the ritual sacrifice of prisoners. The collapse of Classic Maya culture and the abandonment of the cities are now believed to have been caused by a combination of ecological disaster and war. Maya culture survived into the Post Classic (900–1530 CE), though its main centers moved to the north of the Yucatan peninsula where they endured until the Spanish conquest.

The Aztecs of Central Mexico believed that nearly all aspects of life were sacred and were presided over by the gods. They believed that the gods had created and then destroyed four previous worlds. The current world, the fifth age of the sun, would ultimately also be destroyed when one of their gods, Quetzalcoatl, returned.

The Aztecs lived in an unstable region, prone to earthquakes, volcanic activity, and floods. They saw these natural dangers as proof of the instability of the world and a sign of its ultimate doom. In order to postpone this catastrophe, they honored the most important gods and goddesses with human sacrifices. Sacrifice released *Tonalli*, "the warmth of the sun," an energy that the Aztecs believed existed throughout the universe and in all living things. The sacrificial dead gave their energy to the sun, to help it rise and fight against the evil forces in the sky. The Aztec wars of conquest ensured a steady supply of captured warriors to sacrifice to the gods, and slaves were also purchased for the altars. The fifth age closed early for the Aztecs, and it was brought to an end, not by the gods but by the conquistador Hernan Cortés, whom the last Aztec emperor, Moctezuma, mistakenly believed to be the returning Quetzalcoatl.

Right: Tezcatlipoca was one of the most powerful and influential gods in Aztec mythology.

Chalchihuitlicue Goddess of storms

Chalchihuitlicue enjoyed a dual role: she was a goddess of waters, especially stormy waters, and a goddess of youthful beauty. As the wife of the rain god, Tlaloc, she helped him rule the paradise kingdom of Tlalocan. Whereas the Aztecs discerned Tlaloc's presence in the falling rain and in storms, they identified Chalchihuitlicue with the places where that rain collected, linking her to whirlpools, floods and drowning waters. She was seen as a fertility figure with a strong destructive side but the nurturing waters of the womb were also associated with her. This adolescent maiden of spring, whose name meant "Jade Skirt," helped babies and plants to grow. The Aztecs sacrificed children to her and gave her the symbol of a serpent. Because all rivers belonged to her, and had their origins in the heavenly Tlalocan, they also portrayed her as a river, or as a goddess from whom a river flowed.

Chalchihuitlicue created and destroyed the previous age of the world and turned its people into fish.

Cihuacoatl The divine midwife

An earth and mother goddess, Cihuacoatl had many functions. In her positive aspect, she presided over both agriculture and midwifery. When giving birth, women were encouraged to call out her name. This may have been because the goddess purified the birthing process or because Cihuacoatl was a warrior goddess and giving birth was the act of a woman warrior, thus invoking the goddess's name showed the courage of the woman in labor. Male warriors also associated themselves with her and wove the hair of women who had given birth into their shields so that Cihuacoatl would give them additional strength in battle. In her negative aspect, Cihuacoatl was a deeply feared goddess of destruction known as "snake woman", who encouraged young men to sleep with her; after doing so they died. For the Aztecs, the earth was mysterious, dark, and unknowable. Statues of Cihuacoatl were kept in darkness and even her priests did not touch her image.

In Aztec art, Cihuacoatl is shown holding a shield decorated with eagle feathers. Her bare jawbone is often clearly visible, her hungry and destructive mouth gaping.

Cinteotl God of maize

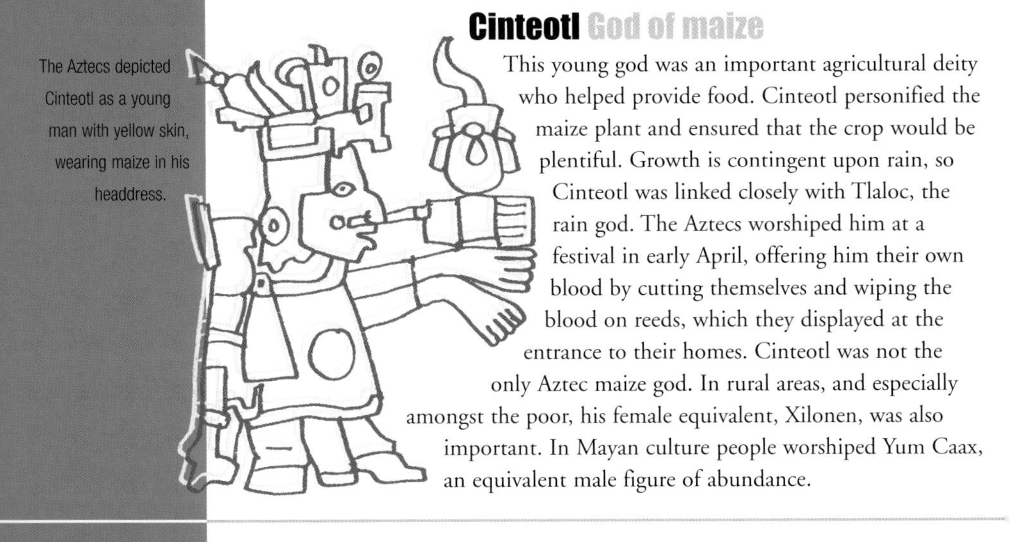

The Aztecs depicted Cinteotl as a young man with yellow skin, wearing maize in his headdress.

This young god was an important agricultural deity who helped provide food. Cinteotl personified the maize plant and ensured that the crop would be plentiful. Growth is contingent upon rain, so Cinteotl was linked closely with Tlaloc, the rain god. The Aztecs worshiped him at a festival in early April, offering him their own blood by cutting themselves and wiping the blood on reeds, which they displayed at the entrance to their homes. Cinteotl was not the only Aztec maize god. In rural areas, and especially amongst the poor, his female equivalent, Xilonen, was also important. In Mayan culture people worshiped Yum Caax, an equivalent male figure of abundance.

Coatlicue Earth goddess

Because Coatlicue conceived Huitzilopochtli without a male partner, early Catholic missionaries linked her to the Virgin Mary.

The mother of the warrior god Huitzilopochtli, Coatlicue was an earth goddess with wide-ranging responsibilities. She helped ensure agricultural prosperity and the efficiency of the Aztec nation's administration. She was also linked to the success of the hunt and the abundance of game. At the main hunting festival, the Aztecs sacrificed a woman to her, and she was also celebrated at the main spring festival, which inaugurated the season of rain, so vital to their crops. Coatlicue means "snake skirt," and the goddess was depicted with a skirt of entwined snakes. She was often pictured with a necklace on which the hearts, hands, and skulls of sacrificed victims hung – hands were considered to be choice cuts for deities to savour. Her place in mythology is particularly important, as she was the mother of Huitzilopochtli. When a ball of feathers fell from the sky on her body and made her pregnant, she was wrongly accused of being promiscuous, and her 400 sons (and in one tradition her daughter Coyolxauhqui) were about to kill her when her warrior son Huitzilopochtli jumped fully armed from her womb and massacred his brothers and sister.

Coyolxauhqui Goddess of the moon

There are conflicting accounts of the death of Coyolxauhqui. In the most sympathetic, Coyolxauhqui warned her mother that her 400 sons were planning to kill her, but when Huitzilopochtli killed his 400 brothers he also slew Coyolxauhqui. When Coatlicue learned of the massacre, she lamented that her daughter had been on the side of good so Huitzilopochtli cut off Coyolxauhqui's head and threw it into the sky, where it continued to live on as the moon. The brothers became the stars. The figure 400 should not be taken literally, the Aztecs used the number to signify "countless." In less sympathetic accounts, Coyolxauhqui took the side of her brothers, but this did not alter her fate. The goddess's name means "bells of gold" and the Aztecs depicted her surrounded by lunar symbols and with bells on her cheeks.

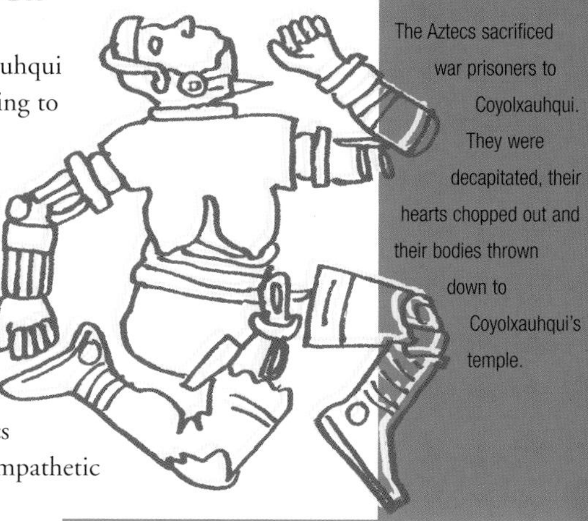

The Aztecs sacrificed war prisoners to Coyolxauhqui. They were decapitated, their hearts chopped out and their bodies thrown down to Coyolxauhqui's temple.

THE BREAD OF LIFE

Maize, the main foodstuff, was central to Mesoamerican life. The Aztecs dried the grain and kept it in store so that it could be used whenever it was wanted. The tortilla is a modern version of Aztec food. The agricultural system was regulated by the Calpulli, a group of families who determined when the essential tasks of sowing and harvesting should take place and how land should be distributed. Farming was not a secular activity: maize was the child of the Earth Goddess and the earth itself was sacred.

Hero twins Maya culture heroes

According to the *Popol Vuh,* the mid-sixteenth century holy book of the Quiché Maya, the twin heroes Hunahpu (or Hun Ajaw) and Xbalanque (or Yax Balaam) were the sons of the maize god who had been sacrificed by the nine dread Lords of Xibalba, the gods of the underworld. After many trials, the twins outwitted the lords of death and avenged their father. Bringing the maize god back to life, they were instrumental in the creation of humankind. Once their work was done, they rose to heaven where they were transformed into the sun and moon. Hunahpu, represented with a black-flecked torso, ruled over humans and was the embodiment of kingship. Xbalanque, who has jaguar ears and patches of jaguar fur on his body, was the ruler of the forest animals. In one myth he slays a giant bird thought to be a representation of the creator god Itzamna, which signifies his mastery of the animal world, and his origin as a god of hunting.

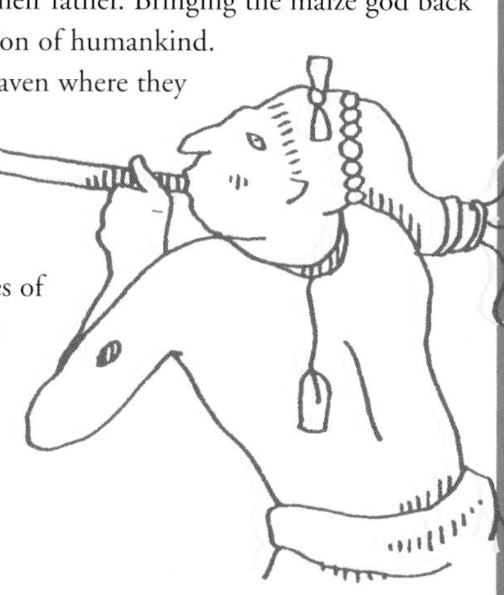

The Hero Twins are Maya culture heroes who are seen as the progenitors of humanity. One twin is in charge of the human world, while his brother is the god of the natural world.

Huitzilopochtli The god of war

Few images of the god survive, as Huitzilopochtli was generally sculpted in wood rather than stone. The Aztecs depicted him with a headdress of hummingbird feathers and in the colors of the bird.

Huitzilopochtli emerged fully armed and fighting from his mother Coatlicue's womb and became the protector of the Aztec nation, a symbol of its might. A courageous figure who may have had his roots in an historical warrior who led the Aztecs to victory, Huitzilopochtli was their supreme god and a central figure in their mythology. In a clear reference to his power and the esteem in which he was held, the Aztecs linked him with the sun at its zenith and with fire. He was honored at the year's main festival, Panquetzaliztli, a festival of dancing during which slaves were killed in mock battles to mark the beginning of the military season. Huitzilopochtli's name meant "hummingbird;" the bird was associated with the idea and practice of sacrifice.

MAYA BLOOD-LETTING AND SACRIFICE

Like the Aztecs, the Maya believed in the mystical power of blood, and offered it to their gods. The most powerful blood was thought to be royal blood, hence the aim of Maya warfare was to capture high-ranking members of the royal houses of rival cities, who would be sacrificed. Another blood offering was performed by the Maya kings and queens themselves, who pierced their own tongues and genitals and let the blood drip onto bark paper, which was then burnt as offerings.

Itzamna "Lizard House"

Itzamna was one of the high gods of the Maya pantheon. He was the creator of both gods and men.

One of the high gods of the Maya pantheon, Itzamna, "Lizard House," was a sky god, and was also credited with the invention of the arts of healing, drawing, writing, and divination. In some traditions, he is said to be the father and creator of humankind and the gods. He is represented as a wise old man with a hooked nose and a flower headdress that falls over his face and also as a giant bird, in whose guise he was fought and defeated by the Hero Twins.

The Maya word "itz" means both "nectar" and "dew." He was the dew of the clouds and the heavens. The Maya priests collected dew for use in their rituals. Itzamna is closely connected to the World Tree, the axis that links together the heavens, earth, and underworld and thus has the power to grant access to the divine world. Another power ascribed to him is to bring people back to life.

Ix Chel "lady rainbow"

The "Lady Rainbow," was the old moon goddess in Maya mythology, in which human activities were associated with phases of the moon. Ix Chel was depicted as an old woman wearing a skirt with crossed bones, and she carried a serpent in her hand. She had an assistant sky serpent, who was believed to carry all of the waters of the heavens in its belly. She was often shown carrying a great jug filled with water, which she overturned to send floods and powerful rainstorms to Earth. Her husband was either the creator god Itzamna or the sun god K'inich Ajaw. Ix Chel had a kinder side and was worshiped as the protector of weavers and women in childbirth.

The goddess Ix Chel was associated with he moon, but was also a goddess of healing, weaving, and childbirth.

K'inich Ajaw Sun-faced lord

In some traditions, the Maya sun god K'inich Ajaw, or Kin Ajaw, was associated with the high creator god, Itzamna, in others he was said to be his son and principal assistant. A powerful figure associated with the institution of divine Maya kingship, K'inich Ajaw is also closely connected with war, human sacrifice, and royal blood-letting rituals. Perfectly preserved murals from the Classic-period Maya city of Bonampak depict warrior-lords performing a blood-sacrifice dance for the sun god. The sun god possessed many aspects related to his journey across the sky by day and through the underworld by night, which were connected with Maya notions of life, death, and rebirth. He was represented in many different guises in Maya art. One of K'inich Ajaw's most common transformations was his aspect as the "Jaguar Sun" of the underworld. His presence was sometimes symbolized by a large mirror during rituals dedicated to him. The Maya believed that the sun's rays could be transformed into centipedes, and these are often represented on his statues, sometimes issuing from his mouth. Skeletal centipedes were also symbols of death and darkness, which were associated with the first rays of the sun as it rose from the underworld at dawn.

The sun god K'inich Ajaw was second only to Itzamna in importance in the Maya pantheon.

Mictlantecuhtli Lord of the realm of the dead

Sometimes Aztec images of Mictlantecuhtli show him with a headdress of owl feathers; the Aztecs associated the owl with death. Sometimes he was given a paper headdress, as paper was offered to the dead.

The underworld was a dark, frightening, and silent place called Mictlan. The Aztecs believed that the vast majority of mortals would endure an afterlife there. Mictlantecuhtli ruled Mictlan, accompanied by his wife Mictlantecichuatl, who had sagging breasts, a skull-face, and a skirt of serpents. Mictlantecuhtli was also depicted as a skeletal figure, often shown grinning, presumably at the likely fate of the onlooker. Mictlantecuhtli had different sides to his personality: on the one hand, he enjoyed the pain of others and in a key myth he tried to trick Quetzalcoatl into remaining in Mictlan indefinitely; yet he could also grant life. The Aztecs used to bury red dogs with their loved ones, to accompany the soul of the dead on its hazardous passage to the underworld as the journey involved various obstacles including a freezing wind of knives that removed bodily flesh from the soul, a giant snake, and a huge lizard. These knives are sometimes shown as Mictlantecuhtli's headdress and, because dogs escorted the soul, he was also the patron of dogs.

Mixcoatl The god of hunting

The Aztecs identified the Milky Way as Mixcoatl.

The Sun and Mother Earth gave birth to 400 stars, but these offspring behaved badly, refusing to be supportive of their parents. To rectify the situation the Sun and Mother Earth created Mixcoatl and four doubles, but Mixcoatl ambushed his star-brothers and defeated them. He was the Aztec god of war and god of hunting, a heroic figure, linked from the start to ambush and violence. He was the first god to make fire (the Aztecs believed the world age in which they lived was created out of fire) and he created war. He may have been the mythological version of an impressive historical warrior. The Aztecs portrayed him wearing striped war paint and carrying his bow and arrows as well as a sacrificial knife. At his October festival, Aztec hunters offered him their own blood.

Quetzalcoatl The feathered serpent

The god Quetzalcoatl was based on an historical figure, a ruler of the ancient Toltec empire who in the eighth century CE united different tribal groups and introduced maize farming. As lord of the winds, Quetzalcoatl drove the clouds that brought rain. A creator figure, he mixed the bones of mortals from the previous age with clay and his own blood to make the first man and woman of the present age of the Aztecs. Quetzalcoatl was also a hero god, believed to be the manifestation of the morning star, which emerged out of a dawn that was red with blood from Quetzalcoatl's fight with his dark twin Xolotl, the monstrous evening star. He was the lord of healing and of healing herbs, the ancestor of the kings, and the patron of craftsmen. He was also closely linked to poetry and culture. He was often painted with a face black with soot, wearing designs that represented the wind.

The Aztecs believed that Quetzalcoatl would return incarnated as a human. When the Spanish conqueror Cortés arrived, they thought the plumes on his helmet were the feathers of Quetzalcoatl and took him for Quetzalcoatl.

THE TEMPLE OF QUETZALCOATL

Carvings of the plumed serpent of Quetzalcoatl adorn the ancient pyramid temple dedicated at him at Teotihuacan, Mexico. The sculptures alternate with images of the god Tlaloc. Built in stages over a period from around 1 CE to around 470 CE, the magnificent building has a large quadrangular court that faces the four points of the compass and a staircase that rises up to the morning sun.

Tezcatlipoca The god of sorcery

In Tezcatlipoca's main sacrifice, a youth was kept in luxury and bliss for a year and then went willingly to his death, where he was decapitated and his heart cut out. Then a new youth was chosen for the following year's ceremony.

Tezcatlipoca, whose name meant "Smoking Mirror," took his name from the glassy black volcanic obsidian stone that Aztec magicians used to look into the future. Tezcatlipoca was the god of sorcery, an adept shape-shifter appearing in different guises to sow conflict in the lives of mortals, whom he frequently mocked. A trickster figure, he often featured in Aztec legend as the adversary of Quetzalcoatl. Yet he was also a creator: he made the first music and, along with Quetzalcoatl, fashioned the first mortals of the current age. He could bring good fortune as well as take it away, and was the patron of an elite cult of fighters, the jaguar warriors. The Aztec kings identified themselves with Tezcatlipoca; to strengthen the connection, every year they dressed a man up to look like the god, then ritually sacrificed him.

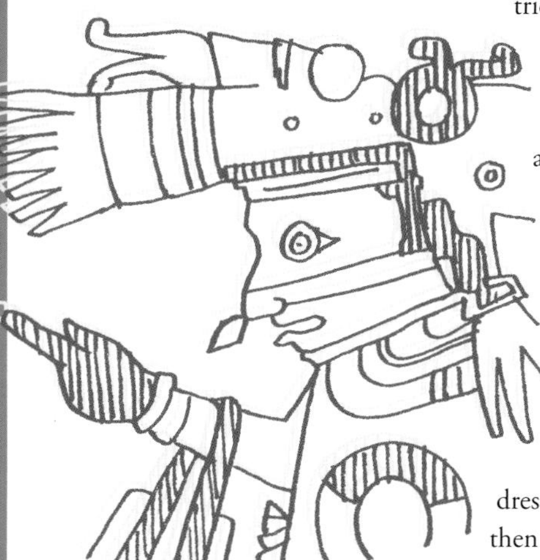

Tlaloc The god of rain

The Aztecs often drew pictures of Tlaloc holding a jug from which he dispensed water. They believed he kept water in giant clay pots on his mountain and that it rained when he broke a pot.

The Aztec Tlaloc, associated with the Maya rain god Chac, was widely worshiped. He was closely associated with the changing cycle of nature since he brought the rain that was so vital to the growth of the agricultural crop. He presided over the paradise of Tlalocan, which was mainly reserved for mortals who had died by his agency, either in the storms and lightning he brought, or from drowning. He had been the supreme god in the third world, an era of rain, now long since gone. The Aztecs believed Tlaloc lived on the mountain where they located Tlalocan, which they believed was the source of water. They envisaged him to be part jaguar, the wild animal they most respected, or with jaguar teeth and goggle eyes, often holding maize in his hand.

Tlazolteotl The goddess of filth

No doubt every therapist would want to get Tlazolteotl on the couch. Known as "the lady of witches" and "two-faced eater of filth" she presided over the things that induced guilt in society, though the Aztecs did not present her in such terms. To them, she had four different aspects that reflected the different phases of the moon. As an adolescent, she was cruel but lovely. As a young woman, she was morally dubious, adventurous and the goddess of gambling. In the following phase she could forgive sin and purify (the early Christian missionaries focused closely on this aspect of her). In her fourth phase she acted as a monster and a thief. Tlazolteotl presided over all sexual behavior. She was the power behind debauchery and evil, inspiring such acts as well as adultery, sexual misconduct, and lust. The Aztecs depicted her as a squatting figure, perhaps defecating, wearing the skin of a human victim as a dress.

The Aztecs employed prostitutes to service the needs of warriors in training. The prostitutes were dedicated to Tlazolteotl and honored her through their work, but it made them unclean, so they were ritually killed and their bodies disposed of in the marshes.

Xipe Totec The flayed god

When a plant begins to sprout, the young shoot splits and breaks out of the seed. The Aztecs understood this process to be self-torture on the part of nature. The agricultural deity Xipe Totec, whose name meant "flayed lord," personified the sprouting or "flayed" seed. The Aztecs also sacrificed to him as a healing god so that he would not inflict plagues or illness upon them. He was the patron of gladiators and at his festival prisoners of war were forced to fight each other; the vanquished were flayed and the victors wore the skins until they rotted. Xipe Totec was depicted in art wearing a human skin.

Xipe Totec's name is related to the Aztec word for "phallus." It is possible that the flayed skin he wore and was so closely associated with referred to the skin of the phallus.

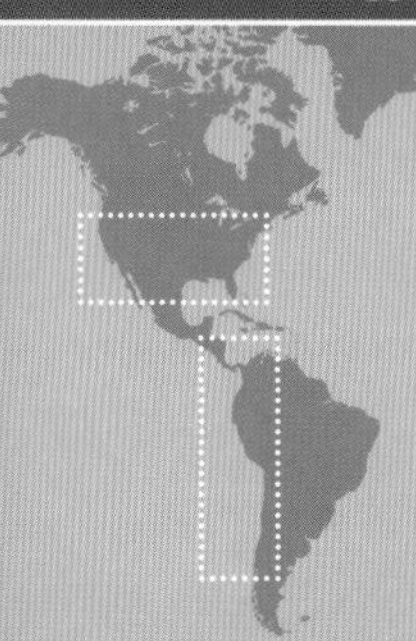

NATIVE NORTH AND SOUTH AMERICAN GODS

The Inca ruled a vast area of the South American continent. This empire included zones of arid coastal desert, the Andean mountains, and lush tropical rainforest. Like the Aztecs, the Inca were the heirs of several urban civilizations. In complete contrast, the native peoples of North America did not develop complex urban cultures. They were for the most part hunter-gatherers who believed that the whole of creation was embued with a divine spirit, and who lived in close harmony with the natural world.

Above: In the north, Native Americans spread from the north west through to the south east of the US. In South America, the Inca empire covered Peru, Bolivia, Ecuador, and most of Chile.

Covering an area now occupied by the modern countries of Ecuador, Peru, Bolivia, and Chile, the Inca empire was the heir of several sophisticated urban cultures, including the Chavin (after 800 BCE), the Moche, and Nazca (200-1000 CE), Tihuanaco (200-1000 CE), the Huari (800-1200 CE), and the Chimu (1200-1400 CE). The empire, which was destroyed by the conquistadors in 1532, was around a century old when the Spanish arrived in the Americas. Because none of the Andean cultures developed a system of writing, what we know about the Inca pantheon and religious practices comes from accounts written after the Spanish conquest, often by unsympathetic Christian commentators.

Like the Aztecs, the Inca were a militaristic people who rose from humble beginnings to rule a vast and diverse empire. They adopted many of the gods of the peoples they conquered, transporting their statues to the Inca capital of Cuzco. They were included as subsidiary divinities in the Inca's own imperial pantheon, which centered around the cult of the sun god Inti, who was believed to be the father of the Inca ruler. Ancestor worship was an important part of Andean religion, and the Inca revered the mummies of their dead rulers, which they housed in their main temple in Cuzco, and paraded in the streets of the capital on festival days much like Christian images are paraded through the streets of Lima today.

The North American landmass, in contrast to Central and South America, remained sparsely populated until the age of European settlement. The Native North American cultures were predominantly hunter-gatherers who exploited a huge variety of ecological niches: the rich coastlines of the northwest, the deserts of the southwest, the woodlands of the Eastern seaboard and the Great Plains of the interior. Although human occupation of the Great Plains dates back to around 12,000 BCE, the nomadic lifestyle that we most associate with the Native North American peoples only came into being in the seventeenth century, once the horse, long extinct in the Americas, had been re-introduced by European settlers. Few of the myths of Native North America concern themselves with the creation of the world by all-powerful gods; rather they deal with the workings of the nature and the cosmos. In North American beliefs, animals and plants, gods and spirits, humans and the elements are all sentient beings who share the same personal nature and are in a reciprocal relationship of interdependence.

Right: The Inca creator god, Viracocha. Many people were sacrificed to this important god, especially children, for whom it was considered a special honor.

Changing Woman Mother of humankind

This Navajo and Apache goddess is said to be the mother of all humans. The Apache also call her the "White Painted Lady." She was born after the first mythic ancestors emerged from the underworld onto the surface of the earth, either alone or in the company of a twin sister.

She became pregnant when touched by the rays of the sun and drops of water and gave birth to twin boys, Enemy Slayer and Born of Water. The twins wanted to know the identity of their father but the Changing Woman refused to tell them. The twins set out on a quest to find their father. When they finally reached the home of the sun, he tested them to see if they were really his sons. Equipped as warriors by their father, the twins returned to earth to slay the monsters and prepare for the arrival of humanity. Changing Woman was a benevolent figure for both the Navajo and the Apache, and was the model for all women. She oversaw the puberty rites of girls.

A Navajo and Apache Mother Goddess from the southwestern United Sates, Changing Woman is a benevolent deity who helps humanity

Inti Lord of the sun

The Inca sun god Inti had several manifestations: as Apu-Inti ("Lord Sun") and Churi-Inti ("Child Son") he was associated with the winter and summer solstices; as Inti-Guauqui ("Brother Sun"), he was the father of the Inca ruler. According to some traditions, he was created by Viracocha on a sacred island in Lake Titicaca. As he rose over the lake, he spoke to the first Inca ruler, the mythical Manco Capac, whom he instructed in his worship. The wandering Inca, led by Manco Capac, were told by Inti to build their capital at a place where a golden rod he had given them would sink into the ground. This would become the city of Cuzco where the Inca built the Coricancha, the main shrine to Inti in the empire. Inti crossed the sky daily and plunged into the Western ocean at night. After swimming under the earth he reappeared each morning. Eclipses were said to be caused by his anger. He was portrayed as a human figure surrounded by a fiery halo of flames and sunbeams. He was central to rituals involving the cultivation of maize, the staple food of the Inca people.

The Inca sun god Inti was said to be the direct ancestor of the Inca ruler. He was worshiped in the Coricancha, the Inca empire's main temple in Cuzco.

Quilla Goddess of the moon

The Inca goddess Quilla was the wife of the sun god Inti and the mother of the Inca queen.

The Inca moon goddess Quilla, or Mama Quilla, was the sister and wife of the sun god Inti. She was worshiped by women and presided over childbirth.

She was held to be the mother of the *Coya*, the Inca queen, just as Inti was believed to be the father of the Inca. She was portrayed as a silver disk with a human face. According to Spanish writers, her cult statue, made of solid silver, was housed alongside Inti's in the Coricancha temple in Cuzco. Quilla was associated with the earth and the dead. With the exception of the planet Venus, the other planets and stars were said to be her attendants. Like the sun, the moon was believed to disappear under the earth when they were not visible.

THE CORICANCHA TEMPLE

This huge complex of buildings was the principal temple of the Inca empire in the capital, Cuzco. Its walls were said to be covered in sheets of gold and silver that caught the rays of the rising sun. It contained the solid gold cult statues of Inti, the sun god, and the mummies or replica mummies of all the former Inca rulers, kept in niches set around the main statue of Inti. The sun god was served by a large number of priests and, he has his own mortal brides, the virgins of the sun.

Tirawahat "Great Father"

Among the Native North American peoples of the Great Plains the Pawnee probably had the most developed cosmology, with an all-powerful creator god known as Tirawahat

Tirawahat or Tirawa, the Pawnee creator god was known as the "Great Power" or the "Great Father." The Pawnee, originally from the area around the Platte River in Nebraska, lived in permanent settlements of earth lodges and engaged in agriculture.They developed one of the most sophisticated cosmological systems of the Native North Americans peoples of the Great Plains. Another exceptional feature of Pawnee religion was that they addressed their creator god directly in their rituals, which sometimes included human sacrifice. Tirawahat first created Shakuru, the sun, and Pah, the moon, to give light to the cosmos, and then set the Evening, Morning, Pole and Death stars to guard the four cardinal points. These subsidiary divinities fought one another, were reconciled and finally created the earth which they populated with humans. But before creation could become truly animate and function, Tirawahat had to intervene directly by performing primordial acts of thinking, gesturing, breathing, lightning and thundering.

Pachacamac God of earthquakes

The god of earthquakes and the underworld was formerly the supreme creator god of a lowland coastal people conquered by the Inca. The Pacific coast of South America is particularly prone to earthquakes, which the god was said to cause by shaking his head. Pachacamac's main sanctuary was located in the coastal site that bears his name near the modern-day Peruvian capital of Lima. He was also believed to send pestilence from his kingdom in the underworld. In some traditions, he was the son of the sun god Inti, and the brother and rival of the gods Viracocha and Manco Capac. His consort was the earth mother Pachamama, with whom he ruled the waters of the underworld. With his daughters, he was the ruler of the sea. He was sometimes represented as a sculpted wooden pillar set atop a pyramid and kept in a darkened chamber.

The Inca god of earthquakes, Pachacamac was originally a creator god of the people of the Pacific coast who was adopted into the Inca pantheon as a secondary deity to Viracocha and Inti.

Viracocha Creator god

The Inca creator god Viracocha first fashioned the earth and sky. He populated the earth with a race of giants. His first creation, however, displeased him, and he decided to destroy it. He turned some of the giants to stone, which explained the giant statues the Inca had seen at the ruined city of Tihuanaco, and the rest he destroyed with fire or water. Viracocha then came back to earth, some say at Tihuanaco or Lake Titicaca for a second attempt. This time he set Inti, the sun, and Quilla, the moon, in the heavens to give his new creation light. He modeled humankind out of clay and painted on their clothes. He gave each their own foodstuffs, songs, customs, even hairstyles, which explained why the peoples of the earth were different. Once he had made the ancestors, he placed them inside caves, mountains and lakes, from which they would emerge to populate the earth. These places became the holy sites of the Andean people. After finishing his work, he departed over the horizon, disappearing over the waves.

The Inca creator god Viracocha created the first humans and gave them their different customs and cultures.

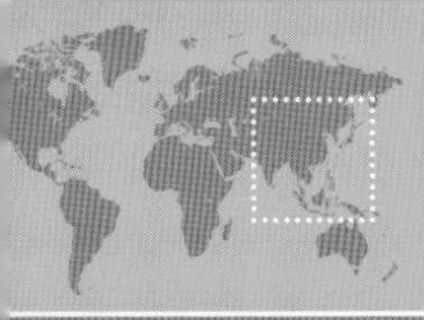

asia

Right: Parvati, the Hindu goddess of beauty and reincarnation of Sati

Below: Golden Buddhas at Fokuangshan Monastery on the Light of Buddha mountain in southern Taiwan.

The Asian continent is the birthplace of several of the world's major religions, both past and present. The Indian subcontinent gave humanity the marvellous complexity of the Hindu tradition, which was disseminated to Southeast Asia. Hinduism itself was the foundation of several subsequent faiths: Sikhism, Jainism and Buddhism. The historical Buddha (born in the sixth century BCE) taught his message of enlightenment in Northern India. However, apart from the island of Sri Lanka, Buddhism is all but extinguished in the subcontinent. In the first centuries of the Christian era it traveled eastward along the Silk Road, establishing itself in Central Asia, Mongolia, Tibet, China, Korea and Japan. In East Asia, Buddhism encountered an interacted with several important local faiths, including Daoism (Taoism) in China and Shinto in Japan.

Asia is the birthplace of two of the world's earliest civilizations: Shang China (c. 1766–1122 BCE) and the Indus Valley culture (c. 2750–1750 BCE) of Northern India, which rival in age and complexity the civilizations of the Near East. Little is known of the earliest religions of Asia, which developed before the invention of writing. With the advent of the written word during the first millennium BCE, in the form of China's ideographic script and the Sanskritic alphabet of India, the names of the gods and the stories of their activities were recorded for posterity. Both cultures worshiped complex polytheistic pantheons of nature gods and goddesses, which over the centuries gradually evolved into symbols of more abstract conceptions of the cosmos and of the nature of human existence.

Above : Dedicated to the Hindu god Shiva, the Sri Meenakshi Temple in Madurai, southern India, is one of the largest and most ancient of its kind.

Not only were the existing religions re-formed and re-invigorated by this process of philosophical speculation, but a powerful new faith emerged in Asia that was to transform the established religions and the world forever. In the sixth century BCE, Gautama Siddharta (c. 563–483 BCE), better known as the Buddha preached his message of compassion and enlightenment in Northern India. Although the Buddha never left India, his message traveled far and wide: South to Sri Lanka, where the oldest tradition of Buddhism (Hinayana, the "Lesser Vehicle") continues to this day, as it does in Thailand; and North and East, where the Mahayana tradition (the "Greater Vehicle") took root in Mongolia, Tibet, China, Korea and Japan. In China, Buddhism encountered and interacted with the two sophisticated and long-established philosophical systems of Daoism (Taoism) and Confucianism, as well as with the folk religion of China with its bureaucratic pantheon subject to the Jade Emperor. In Japan, Tibet, Mongolia and Korea, Buddhism challenged established shamanistic religions, which did not have clearly defined doctrines and were concerned with the worship of nature spirits.

INDIAN GODS

Hinduism has no central doctrine or creed; rather it is a complex web of related mythologies and *yogas* (philosophical systems). Hindus believe in *karma* and reincarnation, and this influences much of the mythology about the gods. The trinity of Brahma, the creator, Vishnu, the preserver, and Shiva, the destroyer, are responsible for the cycles of the world's creation, destruction and re-creation. Although Buddhism is now seen as separate from Hinduism, when it first appeared, it was considered to be one of its unorthodox sects. Buddhism is unique among the world's major religions because it does not believe in an all-powerful divinity that animates and controls creation. Instead it teaches that the goal of life is to attain Buddha nature, the undifferentiated substrate of existence beyond the visible world.

Above: Excavations in the Punjab and Indus valleys (north west India) have revealed the existence of early Hindu culture dating from 3000 BCE.

The three main written ancient sources about Hindu gods and goddesses are the *Vedas* (including the *Rig Veda*, which states that there are 33 gods), the great early epic poems and the *Puranas*. The exact dates of the epic poems, the *Ramayana* and the *Mahabharata*, are unclear. The *Ramayana* was probably composed around 300 BCE, and the *Bhagavad-Gita* (which is an important part of the *Mahabharata*) around a hundred years later. There are eighteen *Puranas*, devoted primarily to the trinity of gods, Brahma, Vishnu and Shiva. Compiled from the fourth century CE onward, the *Puranas* offer a genealogy of the gods and accounts of the creation, destruction and re-creation of the universe. The major gods and goddesses are called *devas*, "the bright ones." They reside in the heavens located in the divine enclosure *Ilvarita*.

Although in its outward forms and popular manifestations, Hinduism is a polytheistic religion, worshiping a multitude of divinities, each with their own sphere of influence, as a philosophy, it is monotheistic, holding that the myriad gods and goddesses are merely aspects of one single transcendent godhead.

Buddhism, the second great faith that appeared in ancient India, is another religion that appears to be polytheistic, in that it venerates a multitude of deified Buddhas or Bodhisattvas (enlightened beings). This, however, is misleading, as the Bodhisattvas are seen to be manifestations of an undifferentiated "Buddha nature," which is not a divine being, but the fundamental nature of the reality that we cannot see because we are blinded by illusion. The founder of Buddhism, Gautama Siddharta (c. 563–483 BCE), remained silent about the gods. Certainly, he did not believe that *devas* controlled human destiny, nor did he advocate their worship. He taught his followers to rely on their own efforts to attain enlightenment, rather than to pray for a god to intervene and aid them. It is likely that he regarded abstract speculation about the existence and nature of god or the possibilities of an afterlife as unhelpful distractions to the spirit of self-reliance that he encouraged his followers to develop.

Right: According to Hindu traditions, Sati immolated herself in shame at the discourtesy of her father toward her husband, Shiva. She was reborn as Parvati.

Agni God of fire

Agni features prominently in the ancient hymns of the *Rig Veda*, as protector of the fire of the domestic hearth and the fire in which sacrifices were offered to the gods. He also acted as a mediator, conveying the offerings of devotees to the gods. As fire, he destroyed darkness but he also consumed his divine parents when he was born. In some accounts he was born as fire when wood was rubbed together, in others as the sparks struck from stone. Like fire, Agni is red and wears clothes of flame. Fire purifies, and worshippers understood Agni to be a purifying deity. He represented the divine will of the universe and appeared in different forms: in the atmosphere he was lightning, in the sky the sun, on earth he was fire. He could grant the prayers of his worshippers and even confer the gift of immortality. For modern Hindus, Agni is mainly approached by men in search of extra virility, and by lovers.

Agni traveled on a ram. It was his vehicle because it was a sacrificial beast and therefore naturally associated with him.

Brahma The Creator

Brahma creates the world in its eternal cycle of creation, preservation and destruction. He is the source of space and time and all that is, the father of gods and people. Brahma is the uncreated creator, known as "the master" and "the architect" by his devotees. He is also "one born from the egg," which refers to the golden egg from which the universe developed. The first of the Hindu trinity of gods that includes Vishnu and Shiva, Brahma has four heads representing the four world ages, the *yugas*, which repeat endlessly in a cycle. According to legend he had a fifth head, but Shiva burned it off with his third eye after Brahma created his daughter Sarasvati and then committed incest with her. Brahma has no major temples and is less frequently worshiped than Vishnu and Shiva. Hindu temples usually have a shrine for him in the northern wall.

A day or *kalpa* of Brahma is 4,320 million years. A night is the same length as a day. A year of Brahma's life is 360 days and nights. Brahma is currently 51 years old.

Buddha The Truthfinder

The historical Buddha was a man who lived in India in the sixth century BCE. In many East Asian traditions, however, the images of deified Buddhas, or Bodhisattvas (enlightened beings), are worshiped in temples and are often associated with divinities from earlier polytheistic pantheons.

The founder of Buddhism, Gautama Siddharta (c. 563–483 BCE) was born in Lumbini in northern India. Of princely birth, legend has it that he was kept sheltered from the woes of the world in the palace, until, during a visit to the outside world, he saw in turn an old man, a sick man and a corpse. Realizing that humanity's lot was old age, disease and death, he resolved to devote his life to freeing man from the cycle of death and rebirth. He first studied and then rejected the leading philosophic teachings of his day; nor could he find the answer he sought in the ascetic practices of the yogis. Determined to achieve enlightenment or die in the attempt, he sat in meditation for three days and nights. On the dawning of the fourth day, he achieved his goal at the age of 35. He continued to preach until his death at Kushinagara at the age of 80 .

Ganesh The god of good luck

Ganesh mythically wrote the *Mahabharata*, from dictation by Krishna. This traditional Hindu epic is eighteen books in length and is the source of many legends. Much of it dates from before the fifth century BCE.

Ganesh is often called "lord of hosts" and "remover of obstacles." The son of Shiva and Parvati, he originated as a demonic figure who could be helpful when flattered and praised but could also induce cause and disaster. He later emerged as a guardian deity, a figure of wisdom, the god of good luck and the embodiment of success and is very popular with Hindus. Ganesh has the head of an elephant, and different explanations are given for this. In one account, his parents coupled like elephants; an alternative account tells how Paravati created from her sweat a boy who would guard her room while she bathed when Shiva was away. Angry that the boy – Ganesh – would not let him enter, Shiva decapitated him, then, to placate Parvati, replaced his head with the first one to hand, that of an elephant. Ganesh is depicted as pink-bodied and large bellied, often one-tusked and riding on a rat.

Indra The god of storms

The king of the gods, according to the *Rig Veda*, Indra originated as a weather god responsible for lightning, thunder and rain. His counterparts are the Greek Zeus or the Roman Jupiter. Probably at a time of conquest, he evolved into a war god, the "lord of hosts," a courageous fighter who destroyed his enemies by tricking them with illusions, then killing them with a hook. His support ensured mortals' victory in battle. He rode a golden chariot and gripped a thunderbolt in one of his four arms, slept with mortal women and killed the serpent of drought. Indra enjoyed a paradise called Svarga, in the company of *apsaras*, divine nymphs who satisfy all pleasures. Indra was a pleasure-seeker who easily succumbed to temptation and was known for stealing the sacrifices meant for other gods; he also enjoyed the hallucinogenic drink *soma*. At some point, Indra declined in importance and was supplanted by Shiva, Vishnu and Krishna, amongst others.

According to the *Ramayana* epic, the gods gave Indra the testicles of a ram after a sage (Gautama) caught him with his wife (Ahalya).

THE FOUR NOBLE TRUTHS AND THE EIGHTFOLD PATH

The Buddha's teachings are contained in the Four Noble Truths:

Life is suffering.
Suffering is born of desire.
Desire can be overcome.
Follow the Eightfold Path.

The Eightfold Path:
Correct thought
Correct action
Correct effort
Correct speech
Correct livelihood
Correct attention
Correct concentration
Correct understanding

Kali The goddess of destruction

Kali is worshiped throughout India as the ferocious and all-devouring aspect of the great goddess Mahadevi. She has a terrifying appearance: an emaciated naked figure whose long thirsting tongue seems to be reaching towards the blood of her victims, wearing a necklace of human heads that drip blood. Her four arms grasp more severed heads and a sword and dagger, although sometimes two of her hands are shown raised in blessing. Known as "the Black One" because she is shrouded in space and night, Kali symbolizes eternal time from which there is no escape, but in which death itself is just a tiny moment in eternity. She also represents the state of non-existence that precedes creation. Kali's destructiveness threatens existence.

Kali's worshippers often offer her goat's blood, but in the distant past human sacrifices were probably made to her. She is especially worshiped in Bengal.

Krishna The divine lover

Scholars debate whether Krishna originated as a heroic warrior or from a dynastic line of leaders in the first millennium BCE.

The divine seducer Krishna is an incarnation of Vishnu. As a child, Krishna performed miracles. He was once in trouble with his mother for eating earth. He denied he had done so, showing her his open mouth, which contained the whole universe, including the planets, stars, lightning, fire and mountains. The bewitching Krishna embodies all the tenderness and sensuousness of love, being playful, irresistible and a figure of contrasts. A god who has come to rid the world of evil and who requires the self-disciplined practice of yoga, he is also a charming stealer of hearts. He especially enjoys the attentions of the celestial milkmaids, or *gopis*, who symbolize the human soul. When making love, he can divide himself into many forms so that he can have multiple lovers simultaneously, each woman feeling that he is hers alone.

Lakshmi The goddess of wealth

Lakshmi and Vishnu enjoy such intimacy that in Hindu iconography they often appear as a single figure, or as an embracing couple.

Also known as Shri, Lakshmi is a popular Hindu goddess, responsible for fertility, wealth and good fortune. A graceful and charming figure, often pictured loyally massaging the feet of her husband Vishnu, Laskshmi is closely identified with domestic harmony. She is sometimes known as “daughter of the ocean of milk,” from a legend in which the gods churned the ocean into milk at the beginning of creation, as they searched for the elixir of immortality. Lakshmi emerged from the milk seated on a lotus flower, decked in lotus garlands and with the complexion of a lotus. The lotus symbolizes fertility and growth, immortality, and refinement and the scene encapsulates the goddess’s refinement and purity. Lakshmi is also represented in one myth of the world’s renewal, which tells how a thousand-petalled golden lotus rises from the cosmic waters to give birth to Brahma. She is worshiped at the autumn festival of Divali, when she is called on to bring good fortune

Mahadevi The great goddess

The supreme goddess Mahadevi is the mother of the world. She is the powerful creative energy of the universe and the essence of reality. The great goddess is intangible, but manifests herself in different aspects including the benign Parvati or the terrible Kali, amongst others. The powerful trinity of Brahma, Vishnu and Shiva were sometimes described by Mahadevi's worshipers as the three legs to her footstool. Her many forms allude to the endless cycle of life and death. Manifested as Parvati, Mahadevi is the consort of Shiva; in this aspect she is his counterpart, the energy of his power. Their divine union created the universe.

The word "devi" was originally a general term for a goddess, later designating a specific goddess, Devi, also known as Mahadevi.

EARTH WITNESS

One of the most popular scenes in Buddhist art depicts the moment when Mara is defeated. The Buddha was not tempted or distracted by Mara, but continued to meditate upon his spiritual perfections. He then touched the earth with his right hand, calling on it to bear witness to his resolve and to the qualities of wisdom and generosity that he had perfected. This is known as the "earth-witness" gesture.

Mara The bringer of death

In Hinduism, Mara is a figure of disease and plague, the personification of death. He is linked closely with Kama. Kama represents desire for life; Mara represents fear of death or, put another way, the desire not to die. In Buddhism, Mara is effectively an equivalent of the western concept of the devil. He is the evil one, the magician of illusion and the great enemy of the Buddha. The meditations of the Buddha resulted in enlightenment, *nirvana,* which freed him from the endless and unsatisfactory cycle of birth, life, death and rebirth. During the Buddha's meditations, Mara tempted him in the form of his three daughters. Temptation is a form of desire. Desire underpins the wheel of existence. If one retains desire and remains attached to worldly concerns, one cannot achieve enlightenment. But the Buddha was unmoved and Mara then hurled a terrible weapon at him, which turned into a garland of flowers around the Buddha's neck.

Mara's daughters are Desire, Discontent and Pleasure. They made an unsuccessful attempt to divert the Buddha from the path to enlightenment.

Parvati Goddess of beauty

Parvati's mythology is a continuation of Sati's and they are often regarded as one goddess.

When the goddess Sati threw herself in the fire and killed herself, she was re-incarnated as Parvati, the beautiful, voluptuous mountain goddess who became Shiva's consort. Shiva is a god of renunciation and Parvati gained his love by sharing his austerity, refusing to eat even dried leaves. When Shiva yielded to her desire, their embrace made the entire earth tremble. She moderates Shiva's behaviour and is sometimes critical of his opium smoking. Parvati personifies the power of Shiva in the cycle of creating, sustaining and destroying the universe.

THE HINDU HEAVENS

The legendary Mount Meru is the center of the Hindu universe. It is the mountain of the gods; their heavens are on and around it. At the summit is the golden city of Brahma; Krishna and Vishnu also have their heavens on the mountain. Krishna's is known as Goloka, the place of cows, and is home to both his devotees and to the milkmaids, *gopis*, who join him in his dancing and lovemaking. To the north of Mount Meru is Svarga, Indra's heaven, which is notable for the apsaras, the celestial nymphs who are the dancers of the gods.

Radha The beautiful milkmaid

In the twelfth century CE the Indian philosopher Nimbarka founded a sect devoted to Radha-Krishna, which established him as an important religious figure.

The *gopis* are milkmaids who accompany Krishna on his moonlit dances of bliss. Radha was his favourite mistress. She was married to a cowherd but was lured by the youthful Krishna into an adulterous relationship. They conducted their affair in the cowherd's village and in the idyllic forest groves that surrounded it. Radha is often understood to be an incarnation of the goddess of beauty, Lakshmi, and came to symbolize perfect devotion and divine love, a strong contrast to Krishna's carnal desires. In one creation account, she was born when Krishna split himself in two. His male half remained Krishna and his female half was Radha, an aspect of the god's power. Radha's desire for Krishna was so strong that her thoughts burned like fire. Their relationship is best understood as a metaphor of the intimacy between the mortal devotee and the divine.

Rama The charming god

Rama originated as a heroic figure who slew an ogre, endured unfair exile, killed the demon king who had imprisoned his wife and then returned from exile to rule over a golden age of peace and prosperity. His exploits and adventures are recounted in the *Ramayana*, the oldest Sanskrit epic, which tells how he won his consort, Sita, and rescued her from the demon-king of Sri Lanka. Rama developed into a god who embodied the ideal qualities of manhood. He lifted and strung the bow of Shiva, an act that required skill and intelligent strength, not brute force. He is a reliable, gentle figure who is widely worshiped as the seventh incarnation of Vishnu.

The heroic deeds of Rama were recorded around 300 BCE, in the *Ramayana*, ascribed to the poet Valmiki.

Sarasvati The goddess of wisdom and science

Sarasvati originated as the personification of an ancient river that had its source in the Himalayan mountains. She has the qualities of a great river: she is active, energetic and strong. Just as a river washes away debris, so Sarasvati washes away sin; she purifies. Rivers also fertilize the ground, and Sarasvati has an important fertility role, providing maternal care to her followers. The functions of the goddess evolved, and she became closely linked to Brahma, as his wife or his daughter, responsible for poetry, music, science and wisdom. She is sometimes said to have been the wife of Vishnu, who, finding her more than he could handle, transferred her to Brahma and another wife to Shiva, retaining only the loyal Lakshmi. In Hindu iconography, she has four hands and is often shown holding a water jug (to symbolize religious ritual) and a book (symbolizing wisdom and learning). She rides on a peacock, playing the lute.

In one creation myth, Sarasvati was born from Brahma. He split into two, a male side and a female side, and then mated with his female side, Sarasvati.

Sati The devoted wife

The word "suttee" derives from "Sati." Suttee is a ritual suicide in which the faithful wife burns herself to death on her husband's funeral pyre. Sati provides the precedent for the act, which has long been illegal, is now very rarely performed and was probably often forced upon the widow.

The goddess Sati loved Shiva and became his first wife, although her sage father Daksa disliked Shiva because of his renunciation of the world, and snubbed him by inviting all the gods except Shiva and Sati to a sacrifice. In outrage, Sati made a sacrificial fire and threw herself in, killing herself. Daksa was killed by demons created by the angry Shiva, whose grief led him to carry Sati's charred body on his head everywhere he went. Concerned that Shiva's sadness was causing discord in the universe, Vishnu followed him, throwing his discus to slice off parts of Sati's corpse, a piece at a time. Shiva ceased grieving when her body had gone, and the places where her body parts fell became sacred places of worship.

Shiva The destroyer

In Hindu mythology, the river Ganges originated in heaven. Shiva divided the river into seven streams so that the water would not catastrophically engulf the earth.

For many Hindus, Shiva is the supreme being. He destroys to create. At the end of a cosmic era, he dances the world out of existence and into a reintegration with the absolute. The Lord of the Dance originated as the ancient storm god Rudra, an important figure in myths from around 1000 BCE. Rudra became Rudra-Shiva, and by the second century BCE the god was known simply as Shiva, one of the trinity that includes Brahma and Vishnu. He is a god of contrasts: closely associated with asceticism, he is also a lover; he is both fierce and calm; he is not intimately involved in the lives of people but is an all-present power who can harm or heal humans. He destroys and yet is a figure of mercy. Shiva is portrayed with matted hair to symbolize his detachment, and with three eyes; the third one, in the center of his forehead, represents the fire of destruction and of wisdom.

LORD OF THE DANCE

Shiva in his most famous aspect, as Nataraja or Lord of the Dance, performing his cosmic dance of annihilation and regeneration. When Shiva dances the dance of Creation, he crushes ignorance under his left foot.

Vishnu The preserver

Vishnu is known as "the Preserver" because he redresses any shifts in the balance between good and evil that may emerge in the universe. He can cross the universe in three steps. Originally a minor deity who personified the energy of the sun, he became increasingly important in Hinduism and is now a far more important figure than Brahma, with whom he forms a trinity of gods along with Shiva. Vishnu embodies goodness and compassion. He is often pictured reclining on the coils of the serpent Ananta, whose head forms a canopy over the god. Ananta is floating on the waters of the ocean, symbolizing a universe not yet created. As the source of the universe, Vishnu rests until he is ready to participate in a new cycle of creation. He is also depicted as a young handsome and four-armed man who holds a conch (a symbol of fertility) a discus (symbolizing speed of thought) and a wheel (representing the endless cycle of the universe).

In each cycle of creation, Vishnu descends to the world in a different incarnation. Krishna and Rama are both incarnations of Vishnu.

Yama The god of the dead

Yama escorts the dead to his realm and looks after them there. His abode is no grim underworld; it is a comfortable and pleasant place known as the "sphere of the ancestors" where children are restored to parents, friend meets friend, and wives are reunited with their husbands. Guarded by monstrous dogs, Yama's sanctuary is filled with supernatural light and free from the ills that characterize life on earth. From being the god of the dead, Yama evolved to be the judge of the dead, determining the fate of the soul, sending it to heaven, to hell, or back to earth for re-birth. Hindus pray to him for long life and for deliverance from death. A negative mythology about Yama has also developed. In Buddhist literature, he is often both the lord of death and of hell. He is most fearsome in Japanese Buddhism, where he is a demon lord, and in Tibet, where he is depicted with bloodshot eyes, dark skin, fangs, wearing skulls on his head and often holding the wheel of existence in his mouth.

Yama became the lord of death because he was the first living being to die.

Above: Ancient China was an isolated region because of its mountains, deserts, and vast seas. Thousands of years passed before overland and sea routes to the west gradually opened up China to Central Asia, India, and Europe.

CHINESE GODS

The earliest city in China, Liang ch'eng chen dates back to 3,500 BCE; however, little is known of the religion of these early inhabitants of China. During the Shang and Chou periods (1766–221 BCE), the Chinese first developed the enduring cult of the ancestors. Rulers and aristocrats were buried in splendid tombs, accompanied by treasures and the bodies of their sacrificed retainers. Between the sixth century BCE and first century CE, the three great philosophies that molded Chinese civilization, Daoism (Taoism), Confucianism and Buddhism, appeared in China. The first two were of Chinese origin, while the third was introduced from India. Popular, or folk, religion combined elements of all three philosophies with the ancient Chinese ancestor cult.

It is difficult to summarize Chinese religion in a short introduction, both because of the geographical extent of China, and its continuous existence over more than five millennia. No other world culture has been so long-lasting, adapting to huge social and political changes, while at the same time retaining its cultural integrity. Evidence from the Shang (1766–1122 BCE) and Chou (1122–221 BCE) periods shows that one of the first elements of Chinese religion was the veneration of the ancestors who were believed to watch over their descendants, rewarding their virtues and punishing their crimes. Along with the ancestors, the Chinese worshiped a pantheon of nature gods. As Chinese civilization grew in size and complexity, the gods also changed, gradually taking on the characteristics of the humans who worshiped them.

During the first millennium BCE, two great philosophies appeared in China: Confucianism and Daoism (or Taoism). The former was an ethical system concerned with the proper ordering of society, which stressed the correct observance of ritual and of ancestor worship as vital to the well-being of the state. The latter was its complete opposite. Daoism did not concern itself with politics or statesmanship. It was based on the concept of the Dao, "the Way," through which men could achieve spiritual enlightenment. During the first millennium CE, different sects of Buddhism were introduced from India and flourished in China. These three philosophic faiths influenced and transformed one another, as well as the folk religion of the people of China, which borrowed ideas and divinities from each, and combined them with its own veneration of the ancestors.

In popular religion, the pantheon of gods is arranged in a hierarchy, much like the complex bureaucracy that governed China for the emperors. The gods report to their superiors, to whom they are accountable, and once a year their actions are reviewed by the supreme god, the Jade Emperor, who demotes or promotes them according to their performance. Worthy mortals who are deified after death join the pantheon, replacing gods who have lost their rank and status as a result of the Jade Emperor's review.

Right: Kuan-yin, the Chinese goddess of mercy or compassion

Eight Immortals

The Eight Immortals are symbols of good fortune and important members of the Daoist pantheon. Originally human, they achieved immortality by following the Dao. They are a popular subject in Chinese art, and their depictions are thought to bring happiness. The traditional list was drawn up in the Sung period (960–1279 BCE). They are: Li Tieguai, who carries a crutch and a magic gourd; Zhong-li Quan, who has a fan; Zhang Guolao, a hermit with magic powers; Lu Dongbin, a hermit who owns a magic sword; Cao Guojiu, a general and brother of Empress Cao; He Xiangu, the only woman, who is able to fly; Han Xiang, who fell from a peach tree and became immortal; and Lan Caihe, who was either a woman or a hermaphrodite.

The Eight Immortals (Pa Hsien) were human sages who achieved immortality through their pursuit of the Dao. They were a popular subject of Chinese art.

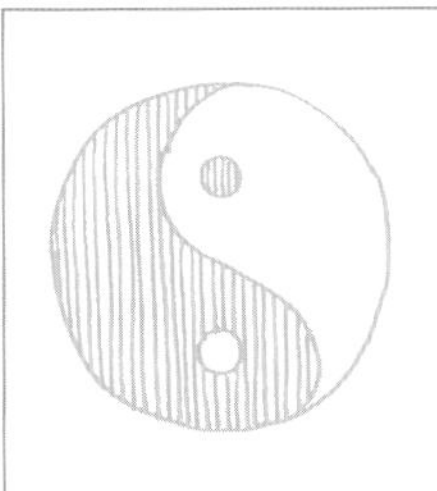

YIN AND YANG

In traditional Chinese thought, the universe is composed of two contrasting but complimentary forces: yin and yang. Yin is the femine, soft, dark principle, and yang is the masculine, hard, light principle. Unlike Western notions of good and evil, yin and yang are not seen as irreconcilable opposites. In the diagram shown here, each half contains the seed of the other. The harmony of the body, state and cosmos is achieved only when the two are balanced.

Huang-ti "The Yellow Emperor"

One of the legendary five emperors of China, Huang-ti (or Huan Di), the "Yellow Emperor" was said to have lived 3,000 years BCE. In one story, Huang-ti achieved immortality at the age of one hundred. He rose to heaven on the back of a dragon, and took his place as one of the rulers of the five cardinal directions. He ruled over the fifth cardinal direction, the center. He was associated with the earth, the imperial color yellow, and the guardianship of the imperial capital.

Huang-ti was also revered as one of the mythical founders of Chinese culture and was credited with the invention of several of the civilized arts, including metallurgy, the use of the compass for navigation, the art of writing, the use of pottery wheels and the breeding of silkworms. Huang-ti represented the ideal Chinese ruler, who is enlightened, wise and cultured.

Huang-ti was one of the five mythical emperors of China. He was revered as one of the founders of Chinese civilization and as the ruler of one of the cardinal points.

Jade Emperor Master of heaven and earth

In Chinese belief, the Jade Emperor became the supreme deity after millions of years of immortal existence, during which he studied and mastered the Dao.

The Jade Emperor, the extensively worshiped Daoist god, is an important figure in Chinese religion. He is an incarnation of the Dao (Tao), the cosmic law of the universe. The Jade Emperor runs the bureaucracy of heaven, delegating authority to other deities. The Chinese pray to lesser gods, hoping that these gods will approach the Jade Emperor, Yu Huang Dadi, the supreme god. The Jade Emperor holds these gods to account; he judges their performance and promotes or demotes them accordingly. He also keeps a record of the achievements of mortals, so that their souls can be judged after death. As the protector of mankind, his earthly responsibilities include the quality of the harvest, the prosperity of mortals and any natural disasters that occur.

Kuan-ti The god of war

Kuan-ti became known as "protector of the empire" in the late sixteenth century.

An important figure in Confucian tradition, Kuan-ti is a war god who *prevents* war. He opposes those who would destroy peace, whether they be internal dissenters or foreign enemies, and is both a figure of righteousness and a guardian of morality. He keeps the frontiers secure. Kuan-ti is based on an historical general from the third century CE, who refused to change sides when captured and was executed as a prisoner of war. Demons and evil spirits fear him greatly. He is the patron of soldiers, the police and state officials and is depicted as a huge figure with a long beard, a ruddy face and red eyes.

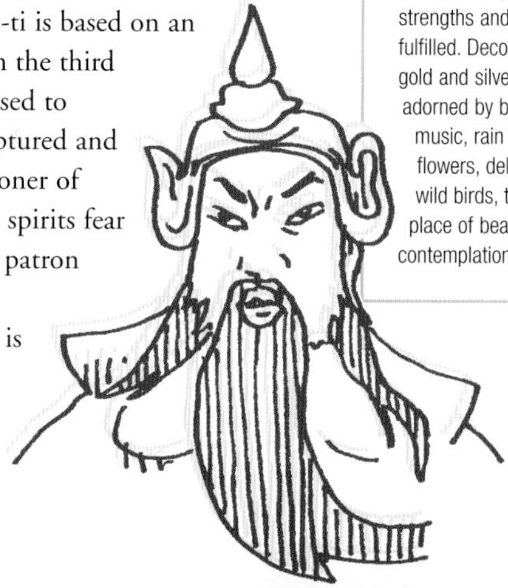

PURE LAND BUDDHISM

In the early Pure Land teachings, the *Visualization Sutra* was a meditation manual to help believers visualize the Pure Land and to be mindful of the Buddha. In it, the Buddha Amitabha is said to have eyes like pure oceans and a circle of light around his head. He has 84,000 physical characteristics, and each one emits a ray of light. The Pure Land itself is a place of calm. There is no suffering and there the reborn will live more like gods than humans, with tremendous powers and strengths and every desire fulfilled. Decorated with jewels, gold and silver, with trees adorned by bells, gentle music, rain that falls as flowers, delightful ponds, wild birds, the Pure Land is a place of beauty and contemplation.

Kuan-yin The goddess of mercy

An important figure in Chinese religion, Kuan-yin originated in India as the male deity Avalokitesvara. He was born from the light of the Buddha Amitabha and was a Bodhisattva (an enlightened being). In China between the seventh and twelfth centuries CE, he became a female and emerged as the mother goddess. Kuan-yin is a Daoist goddess with clear Buddhist roots; her name means "shining over the sound of the world" and she lives on a mountain, where she hears and alleviates human suffering. Women especially pray to her for children, and for their children's health. She watches over childbirth, the different stages of childhood and children's eyesight. Kuan-yin also supports the sick and the senile, protects travellers and looks after the souls of the dead in the underworld. She is a popular figure on Chinese altars, depicted with a halo (showing her enlightened state) and a lotus (showing her purity).

In Tibetan Buddhism Kuan-yin is seen in her male form as Avalokitesvara. Some believe that the current Dali Lama is an incarnation of Avalokitesvara.

Monkey The god of victory

A trickster figure born from a stone egg, Monkey (otherwise known as Sun Wu-kung) was a central character in the Chinese legend "The Journey to the West," in which gods and demons battle with each other. The legend tells how he ruled over the monkeys but wanted to escape death. The Buddha imprisoned him after a dispute with the Jade Emperor, and because he drank the elixir of immortality, he was released from prison so that he could accompany the pilgrim monk Xuanzang on an epic journey. An adept magician, Monkey was a master of the martial arts and extremely well equipped – he had an iron pillar which pounded the Milky Way into existence and magically changed size. He was also a shape-shifter who could travel anywhere in an instant with his cloud-stepping shoes. The legend ends with his triumphal emergence as the god of victory. Monkey grants prosperity and good health to his worshipers.

Monkey is believed to have power over demons and is therefore often prayed to in the hope that he will relieve illnesses they have caused.

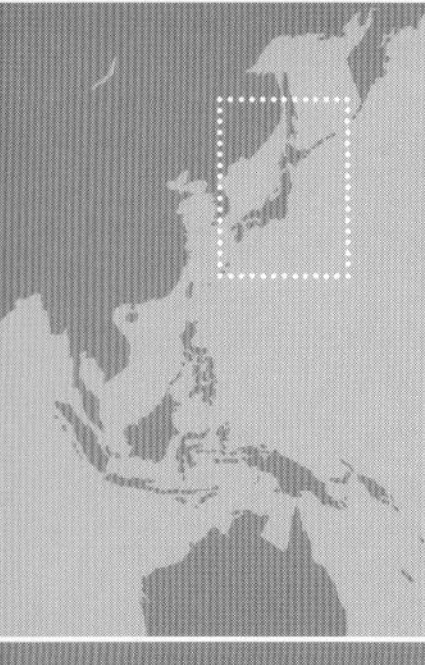

JAPANESE GODS

The island empire of Japan is in several respects similar to the island kingdom of Great Britain. Both are monarchies that once ruled overseas empires; both were civilized by mainland cultures in ancient times – the Britons by the Romans; the Japanese by the Chinese, from whom they acquired their writing system (*kanji* or Chinese characters), their form of imperial government and the Buddhist Religion (in the sixth century CE). The Japanese, however, were not content to be the slavish imitators of their continental neighbors. They quickly adapted the Chinese cultural imports to suit the Japanese temperament and environment. Nor did they abandon their own native beliefs, which they have continued to hold into the modern period.

Above: Shinto and Buddhism continue to play an important role in Japanese life. Marriages are celebrated according to Shinto rites, while funerals follow Buddhist rituals. Japanese traditionally visit Shinto shrines at New Year for luck, and they also place a small Buddhist shrine in their homes dedicated to their ancestors.

The native Japanese religion is called Shinto, and is written with the *kanji* (Chinese characters) *shin* and *do*, which means "the Way of the Gods." Its exact origins are obscure and date back to Japan's prehistory, the Jomon period (10,000–300 BCE). Shinto does not have a single prophetic founder, nor did it have a book of scriptures or doctrines until the composition of the *Kojiki* (*Record of Ancient Matters,* 712 CE*)* and the *Nihon Shoki* (*Chronicle of Japan,* 720 CE*),* which recorded the creation of Japan by the gods and the reigns of its first mythical emperors. Central to Shinto belief is the concept of the *kami,* which is incorrectly translated as "god" and "goddess" in English. *Kami* does include the gods, but also natural objects, such as mountains, volcanoes, trees and rocks, and also human beings, such as the emperor, who was believed to be the direct descendant of the sun goddess, Amaterasu. The Shinto *kami* are worshiped in shrines, in front of which stands a *torii,* a simple wooden gateway painted red. Although Shinto shrines can be quite large, they do not display images of the gods, who are more usually represented by abstract symbols. The sun goddess Amaterasu is symbolized by a mirror.

When the Japanese imported elements of Chinese culture in the sixth century CE, they also adopted the Buddhist religion. Over the centuries, several Buddhist sects were introduced into Japan, including Pure Land Buddhism, which worships the Bodhisattva Amida as its central divinity, and Chinese Ch'an Buddhism, which is now better known as Zen Buddhism, which became the religion of Japan's military elite, the samurai, who ruled the country from 1185 to 1867.

Although Buddhism was the official religion of the empire, Shinto did not disappear. For centuries the two religions were merged, and it was only in 1868, when Japan opened itself to the outside world after two and half centuries of self-imposed isolation, that the two were divided again for political reasons. From 1868 until 1945, when Japan was defeated in World War II, Shinto formed the basis for the Japanese state's imperial cult, which was used as one of the justifications for Japan's wars of aggression.

Right: Right: Okuninushi, the Japanese (or Shinto) God of Medicine.

Amaterasu Sun goddess and imperial ancestor

Born from the left eye of her father Izanagi, the sun goddess Amaterasu is the supreme deity in Japanese mythology. She watches over human affairs and the business of the state. Known as the "heaven-shining great goddess," she fills the heavens with her radiance; when she conceals herself, darkness comes. She foiled her brother Susano-Wo's initial challenge for her kingdom by breaking, eating and spitting out his sword. When he again challenged her supremacy, she retreated to the Rock Cave of Heaven, plunging the earth into darkness while he rampaged. Eight hundred deities tried to entice Amaterasu from her cave with entertainment. The sounds of dancing and laughter drew her to look out, and when a mirror was brought forward, she stepped toward it to view her glorious beauty. The story explains the seasons, as well as periods of fertility and famine.

Until the end of World War II (1939-45), the Japanese state cult taught that the imperial family was descended from the sun goddess, Amaterasu.

Amida The Buddha of measureless light

Known as Amida in Japan and Amitabha in China, the "Buddha of Measureless Light" is also the "Buddha of Eternal Life," the "Buddha of the Western Land" and the "Buddha of the Pure Land" sect that reached Japan via China and Korea around the seventh century CE. The Pure Land sect regards Amida as a saviour, and after the monk Shinran (1173–1261 CE) advocated worship of Amida, the Pure Land sect became the most popular form of Buddhism in Japan. Merely invoking his sacred name "*Namu Amida Butsu*" ("homage to Amida the Buddha") is thought to be enough to guarantee rebirth in Amida's blissful Pure Land paradise. According to the twelfth-century CE monk Honen Shonin, "There shall be no distinction, no regard to male or female, good or bad, exalted or lowly; none shall fail to be in his Land of Purity after having called, with complete faith, on Amida."

According to the teachings of the Pure Land Buddhist sect, chanting the name of Amida is the way to enter the Buddha's western paradise. A concept more reminiscent of the Christian paradise than the Buddhist idea of liberation.

Inari God of rice

As the Shinto god of rice, Inari is an important agricultural deity of Japan. His shrines always contain representations of foxes, which are said to be his messengers.

In Shinto mythology, Inari is the god of rice. As rice is the staple of the Japanese diet and the mainstay of Japanese agriculture, Inari's cult and shrines are found all over the Japanese archipelago. One of the most famous shrines is at Fushimi, near Kyoto, where the entrance is through a tunnel of hundreds of *torii* (usually a shrine has a single *torii*, a simple wooden arch painted red). He is portrayed in various ways: as a human, he is either an old man sitting on a sack of rice, or a young woman with long flowing hair; and as an animal, he is shown as a fox. Statues of foxes are found in front of all his shrines, and foxes are said to be his messengers. His consort is Ukemochi, the food goddess.

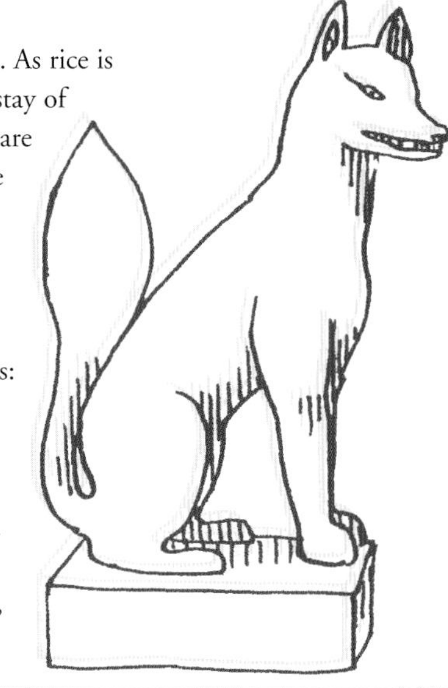

Izanagi and Izanami Parents of the gods

As Izanagi emerged from the underworld, he cleaned himself. From his left eye, he gave birth to the sun goddess, from his right eye, the god of the moon, and from his nose, the god of the sea.

In Japanese mythology, Izanagi and Izanami, the brother-sister spouses, stood on the floating bridge of heaven (presumably a rainbow) and dipped the spear of heaven into the ocean below. When they pulled it out, water dripped from the tip to form the island of Japan, where Izanagi and his wife made their home and produced children. Izanami gave birth to other islands and then to other gods. However, when she gave birth to the god of fire, he so burned her that she lay dying. In rage and grief Izanagi killed her and she went to the underworld where he visited her but fled on seeing her putrid appearance. Izanami became a personification of death and pursued her husband with her attendant demons whom Izanagi managed to pacify. The creator couple became the first married pair to divorce, coming to an agreement at the entrance to the underworld. He took responsibility for the living and she for the dead.

Okuninushi "Great Land Master"

The Shinto god of medicine and magic. He was once the ruler of Yamato, the ancient name for Japan, until Amaterasu replaced him with her own son. He fell in love with Suseri-Hime, the daughter of the storm god Susano-Wo, but the god disapproved of his prospective son-in-law and attempted to kill him to prevent the union. First the god tried to murder him by making him sleep with venomous creatures, but Suseri-Hime protected him; on another occasion, he tried to burn Okuninushi alive, but he was rescued by a mouse who hid him underground while the fire raged. Finally, the lovers eloped and escaped Susano-Wo's anger. At length, the god relented and helped his son-in-law to become the ruler of Yamato.

Okuninushi is the god of medicine and magic, and a legendary ruler of the earth. He married Suser-Hime, the daughter of the storm god, Susano-Wo.

Shichifukujin Seven gods of good fortune

The Shichifukujin are the seven gods of good luck and happiness. Among the group are deities from the Buddhist, Hindu, Shinto and Daoist pantheons. The seven travel in a magical boat filled with treasure and magical items, including a hat that makes its wearer invisible and a purse that is always full of money. They are: Daikoku, the god of wealth and agriculture, who grants the wishes of mortals with his magic mallet; Daikoku's son Ebisu, is a fisherman and the Shinto god of wealth, who is shown holding a fish; Benten, who rides a dragon, is the goddess of love and happy marriages; Bishamon, a Hindu god of war, represents benevolent authority; he is shown dressed as warrior and carries a pagoda and a lance, symbols of his dual role as a religious teacher and a soldier; Fukurokuju is the god of longevity; Jurojin, another god of long life grants a happy old age; and finally, Hotei is a fat, jolly monk who is often shown seated on a large sack, which represents material contentment.

The Shichifukujin are drawn from the Hindu, Daoist, Shinto, and Buddhist pantheons. They represent all human happiness and good fortune.

Susano-Wo God of storms

In legend, Susano-Wo prepared for battle with an eight-headed snake by tricking it into drinking from eight vats of intoxicating distilled alcohol.

Susano-Wo was the wicked storm god, personifying the rage of the sea and the fury of the weather; in some traditions he was also a god of plague. He annoyed his sister, the sun goddess Amaterasu, by defecating in her palace and then challenged her for supremacy, claiming victory despite her protests and rampaging over the earth. He threw a flayed colt into the room where the sun goddess was weaving clothes for the gods, forcing her to withdraw to a cave, leaving the earth in darkness until the other gods encouraged her from her shelter. Susano-Wo found himself tried by a council of gods, who pulled out his fingernails and threw him out of heaven, whereupon he became god of the underworld.

In legend, Susano-Wo killed an eight-bodied snake and pulled a sword from its body. This sword was called Kusanagi no Tsurugi and formed an important part of the insignia of the Japanese imperial family.

THE WEDDED ROCKS AND THE SHRINE OF ISE

The Meoto Iwa or Wedded Rocks are an important place of pilgrimage in Shinto religion and it is considered auspicious for young lovers to visit them. Located in Ise Bay, a few miles from the sacred shrine of Ise, the Wedded Rocks consist of two large rocks linked by straw ropes that symbolize the sacredness of the site. In Shinto belief, natural objects possess a spirit, or kami. The rocks represent and contain the primal couple in Japan's mythical history, the husband-and-wife deities Izanagi and Izanami. The larger of the rocks is Izanagi, the smaller Izanami. The ropes that link them, which are replaced each year, are not used as a bridge. Upon the larger rock stands a torii, the gateway that marks the entrance to a Shinto shrine and indicates the presence of a kami. The torii also alludes to the mythical tradition in which the goddess Amaterasu was enticed beyond the entrance of the Rock Cave of Heaven in which she hid from her brother Susano-Wo.

The Ise Shrine receives over five million visitors each year. It consists of two separate complexes separated by the Isuzu river and linked by a bridge. The outer shrine is dedicated to Toyouke, the god of the harvest, and the inner shrine to Amaterasu. The inner shrine is believed to house not just the goddess herself, but also her most treasured possession, her mirror, which is an important symbol of her divine authority.

The shrines themselves are thatched huts built from cypress wood, and are completely rebuilt every twenty years so that they remain spiritually pure. They were last rebuilt in 1993.

Tengu Long-nosed spirits

Thought to be of Chinese origin, the tengu ("heavenly dogs") are tree-dwelling, telepathic mountain spirits believed to have supernatural powers. They haunt the lonely places of the earth, and they are well known for playing tricks on human travelers. Like the fairies of the Western tradition, they are not evil but are just as capable of helping as they are as hindering the humans they encounter. They are part human and part bird, and are born from giants eggs. Their most distinguishing feature is an outsized phallic nose. They are shape-shifters and have cloaks made of leaves or feathers that enable their wearers to become invisible. A legend about the tragic hero Minamoto no Yoshitsune tells of his encounter with the tengu as a child. Pursued by his enemies, Yoshitsune hid on Mount Kurama. The tengu took care of him, and their king, Sojobo, taught him the martial arts and strategy. However, the cheeky Yoshitsune stole his teacher's cloak of invisibility and played tricks on his benefactor. A tengu could assume a human shape, or transform itself into a badger or fox, but its shadow always betrayed its true nature.

The Tengu are mountain spirits who are half man, half bird. Although not evil, they do not necessarily help the humans they meet. Like the fairy folk of the Western tradition, they are apt to play tricks on mortals.

africa and caribbean

Right: Ogun, the Yoruba god of iron

Below: Traditional celebration in Shakaland, a Zulu village in Natal.

There are literally hundreds of Native African religious traditions. Generally, these are polytheistic and hierarchical, with a supreme being presiding over a number of lesser divinities. Anthropologists often consider the religions to have monotheistic elements, as the authority of the lesser gods stems from a supreme creator. Unlike the subordinate divinities, who concern themselves with human events, often mediating between mortals and the supreme god, the creator remains distant. A similar pattern of belief, with a distant supreme god and intermediary deities, was also established in the Afro-Caribbean religions of the New World.

African cultures have a range of myths that explain why their supreme god is distant. These myths are usually stories in which the god separates from the world and its first people, both his creation, and they explain the origins of sickness, suffering and death. Typically, because the supreme god is often believed to live in or beyond the sky, these myths also involve a distancing between earth and sky. For the Nuer of Sudan, for example, a rope originally linked earth and heaven, and the first people were free to clamber down the rope to collect food until a trickster hyena mischievously cut the rope. God remains in the sky, humankind on earth. In a Yoruba myth, Olorun distanced himself as a punishment for people greedily taking too much food from heaven. He remains aloof, having delegated his authority over the world to the lesser Yoruba divinities called *orisha*. Whereas Olorun is impersonal and distant, the *orisha* are locally involved and active.

Above: Voodoo dolls are used as mediums to carry messages to the spirit world. Dolls stuck with pins are not a reality in voodoo; just a commodity in tourist shops.

The Yoruba offer gifts to the *orisha* in the hope that they will take them to Olorun, and the *orisha* also play an important role in ritual ceremonies that re-enact the events of sacred myth, in divination when people learn sacred knowledge, and in ritual possession, when a medium is filled with the spirit of an *orisha*. Attitudes toward possession vary markedly. For the Dinka, unless the possessed person is a professional diviner or a prophetic figure, possession is a sign of an illness that must be cured; for the Yoruba, it shows a close relationship between individual and *orisha* that strengthens links with the god involved, who acts as a guardian spirit.

A closely related group of religions, which includes Voodoo in Haiti, Santería in Cuba and Candomblé in Brazil, emerged in the Caribbean and the Americas with the arrival of large numbers of slaves from West Africa between the seventeenth and nineteenth centuries. Once in the New World the newcomers were exposed to the new beliefs, which they combined with their own, creating synchretic faiths that combined African, Native American and Roman Catholic elements.

iNkosi yeZulu The lord of the sky

For the Zulus, when the clouds are close to the earth then iNkosi is also close and extra care should be taken in the way one lives one's life.

For South African Zulu, iNkosi is the lord of the sky and the personification of heaven. He is often wrongly identified with Unkulunkulu ("the Old One") as Christian missionaries confused the first man, the great ancestor of the Zulu, with their supreme deity. Known as "lord of lords"and "father of all things," iNkosi is a powerful and wise god who lives in the sky. The Milky Way is the entrance to his home. He manifests himself in thunder, which may be playful or angry according to his temper, and in different forms of lightning. He makes the rain, which the ancestral spirits who act as his intermediaries distribute according to their mood. These spirits also keep him informed about events in the human world.

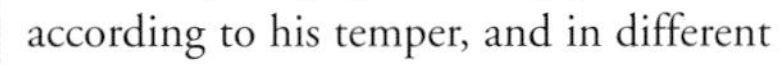

THE CHURCH IN AFRICA

Many African Church movements combine African rituals with Christian beliefs or have separated from the established missionary movements in Africa. These are known as Independent Churches. In South Africa, one of the most important Independent Churches is the Nazareth church founded in 1913 by Isaiah Shembe, the Zulu prophet, healer and preacher. His monotheistic movement stressed the importance of Old Testament laws and strict Sabbath observance. It also advocated honoring, rather than worshiping, ancestors. By the time of his death in 1935, Isaiah Shembe had become a prominent figure in Zulu culture and was regarded by many of his followers as the Black Messiah, the Holy Spirit sent to save the Zulu people.

iNkosikazi yeZulu The goddess of maidens

The Zulus envisage iNkosikazi yeZulu to be a goddess who smiles often, so showing her beautiful teeth, unless she is communicating bad news to mortal females about, for example, an imminent lack of rain.

The virginal princess daughter of iNkosi yeZulu never ages. A fertility figure, she helps to ensure that the rains are bountiful by negotiating with her father. She is good, beautiful and trustworthy and appears to mortals in a mist in the morning, either naked or dressed in white. If a man sees her, he will go blind, for it is not he that the goddess will have come to see – iNkosikazi is the special friend of girls and maidens. She advises them on their choice of husband, helps them learn how to cook and work the fields and is associated with their fertility. She is identified with the rainbow, which is a sign of her glory and is also associated with females rather than males.

Kwoth The supreme god

Kwoth, the supreme god of the Nuer people of southern Sudan, is an intangible figure. His name closely relates to the idea of breathing, of spirit. He is in the air, he is everywhere, like the wind; he is especially in the sky. The great creator manifests himself in rainfall and in lightning and the rainbow is his necklace. Kwoth is both the sustainer of life and the bringer of death: the lightning strikes and whirlwinds that remove mortals from this earth and take them to live as spirits in the sky are his work. He punishes wrongdoers and maintains balance in the affairs of individuals – Kwoth takes away good things from people who are over-eager to celebrate their fortune.

In Nuer mythology, the earth separated from the sky when a naughty hyena that Kwoth had ordered should be guarded, cut the rope that led to earth from the tree of creation in the sky.

Nhialic The supreme god

Although Nhialic, the supreme god of the Dinka people of southern Sudan, resides in the sky, he is not the sky. He presides over the lives of the Dinka and brings the rains that are so essential to existence. He also punishes the wicked and decides the destiny of his people. Yet he is ultimately a distant figure and his separation from the Dinka is the subject of important myths. Originally the first man and the first woman enjoyed a close relationship with Nhialic; a rope connected the earth to the sky and humans travelled freely between the two. But when the first woman was grinding millet, she raised her pestle too high and it hit the sky. Offended, Nhialic sent a bird to cut the rope, an act that separated him from his creation. It also meant that sickness and death entered the world. Nhialic later prevented mortals from rediscovering the bliss and ease they had previously enjoyed. His relationship with the Dinka is complicated: on the one hand, he lets his will become known through the voices of prophetic figures who have appeared throughout history, and the Dinka pray to him to ease suffering and illness; on the other hand, he remains distant and removed. The Dinka say that they do not fully understand him, in the same way that a child does not always understand a parent.

In Dinka mythology, the first mountains and rivers were created and placed as obstacles by Nhialic to prevent the ancestors of the Dinka from finding a drought-free paradise.

Nyame The supreme god

For the Akan, the soul is one's god-given destiny, which after death goes to the land of the ancestors.

The supreme god of the Akan, who live mainly in southern Ghana, resides in the sky. All-powerful and all-seeing, he is a provider and a sustainer of life. He carved the world into existence and then withdrew from his creation. Nyame has divine messengers, lesser deities called *obosom*, who answer prayers addressed to him. The *obosom* are capricious figures, able to cure illness and solve the crises of mortals, but also capable of destructiveness.

Ogun The god of iron

In Yoruba legend, Ogun once massacred a group of people who were ritually worshipping him in silence, mistaking their silence for disrespect.

For the Yoruba of Nigeria, chaos predated creation, in the form of a marshy wasteland. The god Ogun climbed down into the chaos on a spider's thread and cleared the marshes so that other gods could follow him. The wasteland suited Ogun, because he could hunt there. He is the god of war and the god of hunters. He gave the finishing touches to all of creation and brought order to the marshes. In doing so, he paved the way for civilization, which is, to an extent, dependent on the commodity iron, so Ogun is the god of iron and the patron of anyone who works with iron or with steel, especially smiths and barbers, who make regular sacrifices to him. As a war god, Ogun is a destructive figure, but his destructiveness is logical and necessary for order. Ogun is extensively worshiped as a provider who brings both wealth and victory.

Olorun The supreme god

Olorun is also known as Olodumare, which is an older name for him. "Olorun" means "owner of the sky."

The supreme god Olorun heads the hierarchical Yoruba pantheon of Nigeria. He is served by a group of lesser deities known collectively as the *orisha*. Several scholars have argued that Yoruba religion was influenced by the monotheistic religions of Christianity and Islam, and that the *orisha* are manifestations of Olorun rather than separate deities in their own right. As individual gods, however, they have worshippers whose offerings link them to Olorun, who is more distant and less approachable. Although Olorun fashions the destiny of each individual, he does not intervene in the affairs of mortals. Olorun did not create the world. He ordered one of his sons (Obatala) to do so, and also later to create people, although it was a different son (Odudua) who carried out his orders. Then Olorun breathed life into the first people, fashioned out of clay. Olorun is the judge of the dead: the good enjoy a happy afterlife and are reincarnated; the bad endure an endless afterlife in a place of misery and are never reborn.

Oya The goddess of female charm

One of the *orisha* of the Yoruba people of Nigeria, Oya is a weather goddess. Like a storm, she can be unpredictable and her anger is fierce; she appears in the winds, especially gales and tornadoes, in the river Niger, and also manifests herself as lightning. Sometimes she chooses to remain hidden and invisible. Oya is a patron of women, especially of female intelligence and charm; she protects women traders in the marketplace. Oya is also a goddess of the forest, ensuring that there are wild animals to be hunted. She is closely associated with the hunt, and can take either side in it, sometimes turning herself into the beast that is being pursued. She also appears to mortals as a wild buffalo.

Oya is the consort of Ogun, the god of war and thunder.

Shango God of thunder and lightning

The Yoruba god of thunder and lightning is widely worshiped in West Africa. Also known as Xango, Sango and Sagoe, Shango was mythically a ruthless and much-feared king of Oyo, the ancient Yoruba capital. He had a charm that gave him power over lightning, which he used to protect his kingdom from both external and internal enemies. Fearing that his powers were waning, he climbed a hill outside the city to test them but in doing so he inadvertently cast lightning upon his own palace, killing his family. He abdicated and then hanged himself. According to his followers, Shango was deified after his death and continues to rule as an orisha. He creates thunder by throwing "thunderstones" to the earth and punishes fraud and dishonesty by hurling lightning at wrongdoers. Rams are held sacred to him and he is often depicted with horns.

Shango's followers often dance holding wooden staffs that have a double axe motif at the head. These represent Shango's "thunderstones" and are a symbol of his potential destructiveness.

AFRO-CARIBBEAN GODS

Between the sixteenth and nineteenth centuries, millions of Africans were transported from their homes, principally in West Africa, to the New World, where they were used to supplement the falling Native American workforce decimated by war, physical abuse and epidemic diseases. They brought with them their own gods and myths, which over time they combined with Native American and Christian beliefs to create vibrant religions that have not only endured but prospered, finding new homes in the U.S. and Europe. The best known of these religions originated in Haiti, and is popularly known in the West as Voodoo, although in Haiti itself, it is known simply as "serving the spirits."

Above: Voodoo is today the main religion of Haiti. The idea that voodoo is superstition, magic, or devil worship does not diminish its importance to Haitian culture.

Haitian Voodoo, or Vodou in creole, has roots in the religions of several African peoples, including the Yoruba of Nigeria, the Fon of Benin and the Kongo of Zaire. Once they arrived in the New World, the slaves were often forbidden to speak their own languages or worship their own gods, because their European masters were afraid that to let them do so might encourage them to unite and rise up in revolt. The slaves were often forcibly converted to Christianity, in order to subjugate them further, and alienate them from their original beliefs.

Despite all these repressive measures, the slaves on the Caribbean island of Hispaniola rebelled and fought a protracted war of independence, which culminated in the creation of the first Afro-American republic in the New World in 1804. Terrified by this first Black victory, the Europeans gave the new republic and its religion a bad press. Sensationalistic stories of demonic possession, sorcery and zombies still fill the popular imagination when the word Voodoo is mentioned. Although these ideas have more to do with the fevered imaginations of Hollywood producers than the reality of Haitian Voodoo practices.

The Voodoo pantheon is headed by an all-powerful creator god, known as Bondyé (from the "*Bon Dieu*," the French name for the Christian god). Like several of the African supreme gods, however, he does not enter into direct communication with mortals, nor does he take much interest in their affairs. These he entrusts to the care of lesser gods and goddesses, the *Iwa*, who are very similar in role to the Yoruba *orisha*. Several of the Voodoo *Iwa* share the same names as Yoruba *orisha*, while others are of Native American origin or are closely associated with Roman Catholic saints. The *Iwa* are subdivided into several *nanchon* (nations), which have different characteristics and functions. The *Iwa* will manifest themselves during Voodoo religious ceremonies by taking possession of certain of their devotees, who will perform the songs, dances and other actions associated with the particular spirit manifested. As in Yoruba possession by an *orisha*, this is considered to be a blessing by Voodoo devotees and priests, and shows the exceptional favor of an *Iwa* for a devotee.

Right: Ghede – the Voodoo god of the dead also known as Baron Samedi.

Agwé God of the sea

An important member of the *Rada Iwa nanchon*, Agwé is the Haitian god of the oceans, wind, thunder, boats and seashells. He is said to live on an underwater island with his consort, La Sirène (the mermaid), the sea aspect of the goddess of beauty, Erzulie. Agwé is represented as a large sailboat with the word "*Immamou*" painted on its side. In rituals held in his honor, the priest will offer the sacrifice of a blue-colored ram. He is associated with the Roman Catholic Saint Ulrique.

Haitian god of the sea and navigation, he is portrayed as a large sail boat.

THE *IWA NANCHON*

The Voodoo *Iwa* are divided into several *nanchon* (nations), which are thought to be of different origin. The *Rada Iwa* are associated with African gods and carry out their functions, such as regulating the weather or patronizing certain human activities. As such they are close in nature and role to the Yoruba *orisha* after whom they are often named. The *Petro Iwa* are thought to be of Native American origin are far more disruptive and dangerous entities. A third *nanchon* are the Ghede, the nation of the dead, who are ruled by Papa Ghede, also known as Baron Samedi. Certain *Iwa* have a foot in both camps, so to speak, by having both *Rada* and *Petro* manifestations.

Damballah Good serpent of the sky

A powerful serpent god of the *Rada Iwa nanchon*, Damballah is the patron deity of rains, streams and rivers. Although he is a sky god, he is often found in bodies of water, and special pools (*bassins*) are built to accommodate him. He also sometimes settles down to rest in the branches of trees. Originally a Fon deity from Benin, in Haiti, Damballah is also known as Bondyé (from the French Bon Dieu, the name of the Christian god). As the cosmic serpent, he represents the creative life-force of the universe. His consort is the rainbow snake, Ayida Wedo. Together they give birth to the world and their symbol is an egg, which is also their principal offering. They represent the sexual totality, both male and female, self-sufficient and whole. When alone, Danballah is portrayed either as a rainbow-colored serpent forming an arc across the sky, or as a giant snake coiled around the earth and devouring his own tail like the Greek *Omphalos.* Danballah is seen as an ancient, venerable and benevolent paternal deity. However, he rarely concerns himself with the affairs of the human world, which he leave to less senior *Iwa.*

One of the high gods of the Voodoo pantheon, Damballah does not concern himself with mortal affairs, which he leaves to lesser divinities.

Erzulie Goddess of beauty

The Haitian Voodoo Aphrodite-Venus, Erzulie is known for her beauty and her love of luxury.

The great goddess of Voodoo, Erzulie represents the feminine principle in all its complexity and variety. In some manifestations, she has the attributes of the Graeco-Roman goddess of love, Venus-Aphrodite. She is usually shown in elaborate dress and has many lovers, including, the sea god Agwé, the sky serpent Danballah and the warrior god Ogoun. When she possesses a devotee, her first act is to perform an elaborate toilette. She is provided with makeup, fine clothes and expensive perfume. She has both *Rada* and *Petro* aspects. As Erzulie Danto, she is the dominating but homely matriarch, identified with the Black Madonna; as Erzulie Freda, she is the dizzy starlet of the *lwa*, who loves luxurious living; as the *Petro Iwa* Erzulie Ge-Rouge (red eyes), her love and charms can turn into shrieks of fury and hatred.

Ghede God of the dead

The archetypal Voodoo god, Papa Ghede, also known as Baron Samedi, sports a top hat and tail coat, wears dark glasses and carries a walking stick.

Thought to be another Voodoo god of African origin, Ghede is the *lwa* of the dead. While Legba is the sun lord of the crossroads between heaven and earth, Ghede is his dark brother, who stands at the crossroads between life and death. He is also the bawdy, ribald joker, the great leveler, whose appearances are usually welcomed for the hilarity they bring. An inveterate prankster, he is sometimes the uninvited guest at rituals held for other *Iwa*. Ghede knows that sexuality is an inevitable and central part of life. He takes great pleasure in making fun of those who are uncomfortable with their sexuality.

As Baron Samedi (Baron Saturday), his outfit is a top hat and black tail coat, set off with dark glasses. He carries a cane in one hand and a big cigar in the other. More suave than Ghede, the Baron is jovial and jocular but also much more sinister. He is the Lord of the Dead who governs all the occult forces of sorcery and necromancy. His consort is Brigitte, who is named after the Catholic St. Bridget. She is the guardian of graves, a powerful magical *lwa* of cemeteries as the consort of Baron Cimetière (Baron Cemetery), yet another manifestation of Baron Samedi.

Legba God of the crossroads

Originally a Fon solar god and creator of the world, Legba was both male and female, an incarnation of the principle of life. In Haiti, he stands at the crossroads between the world of gods and men, as the guardian of the "Grand Chemin," the great path leading to the divine realm. Without his consent there can be no communication between the *lwa* and mortals and he is the source of all rituals. He also rules the four cardinal points of the compass.

Represented as a cranky, but loveable old man who walks with a stick, Legba seems to be nearing the end of his life. His colors are red and white, and his offerings include grilled chicken, sweet potatoes and tobacco. He is paired with the darker *lwa* Carrefour or Kalfu, the trickster Lunar lord of the crossroads, who rules the points in between the four cardinal points.

The guardian of the "Grand Chemin" between heaven and earth Legba is represented as an aged man standing by a gateway.

VOODOO MAGIC AND SORCERY

Voodoo is the heir of several African traditions of white and black magic that hold that misfortune and illness can be caused by the ill-will of an *lwa*, or of a *boko*, a black magician. It is one of the functions of Voodoo priests to manufacture charms to protect a devotee from sorcery, or to intercede with the *lwa* to discover what can be done to assuage their anger.

Ogoun God of warriors

Ogun is the Yoruba god or *orisha* of iron and blacksmiths. In Haitian Voodoo, as the *lwa* Ogoun, he is an original figure who has many different functions and identities. He is at once a statesman and diplomat, a magician and healer, as well as a warrior. He is a symbol of many forms of political and military authority and power. He is sometimes portrayed as a mortally wounded soldier who is carried by two companions in a pose reminiscent of the Deposition of Christ from the cross. Sharing many attributes with the Graeco-Roman god Ares-Mars, Ogoun is associated with the Roman Catholic St James, the apostle, and sometimes with the warrior, St George, who slayed the dragon. In his warrior aspect he carries a machete or a sword, but in his aspect of Ogoun Feray (from the French fer or ferraille, iron and iron-ware) he is also the patron of the smith's peaceful products, such as the farmer's hoe and the doctor's scalpel. Along with Damballah and Agwé, Ogoun is married to the promiscuous Erzulie. His favorite offerings are tobacco and rum.

Ogoun is the Haitian version of the Yoruba god Ogun, patron of smiths. In Voodoo belief, he is associated with war and warriors.

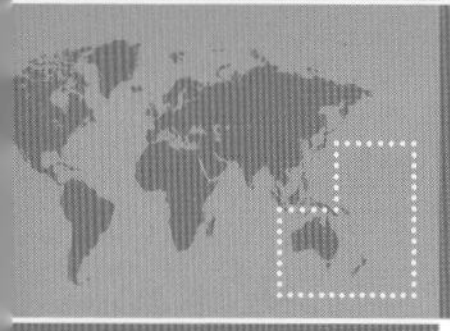

oceania

Right: Papa and Rangi, creators of the six major gods in Maori mythology.

Below: Basalt statues of human figures line an ahu (stone platform) at Tongariki, on the coast of Chile's Easter Island.

Oceania encompasses the continent of Australia and the many thousand islands of the Pacific known as Polynesia, Melanesia and Micronesia. While the Australian continent was settled some 65,000 years BCE, the islands of the Pacific were settled much more recently – many in the last 1,000 years. Australia, apart from its northeastern coastal fringes, is a vast parched desert, and the myths of its peoples are deeply concerned with assuring supplies of food and water. The island peoples of the Pacific, in contrast, lived in much kinder environments than the Australian Aborigines. They developed settled patterns of life, dependent on agriculture and animal husbandry, as well as fishing and hunting, which freed them from some of the uncertainties of nature.

Above: An Aboriginal rock painting shows spirits from the "dreamtime" – the period of creation in Aboriginal mythology.

The Aboriginal peoples of Australia have made their home in one of the most difficult environments on earth, the Australian desert outback, where food and water are scarce during the dry season, and where the rainy season brings sudden and devastating floods. Under these extreme conditions, they have not developed settled patterns of life dependent on agriculture and animal husbandry but many have continued to the present day to live as hunter-gatherers in close harmony with the land and the plants and animals that sustain them. Their religious beliefs are intimately associated with their rights to use the land and its resources, which has brought them into conflict with the white inhabitants of Australia, who have a very different understanding of land rights. One of the central Aboriginal concepts is the period known as "Dreamtime," a primeval age of the earth, when the ancestral spirits appeared and created the natural and human worlds. These ancestral spirits, such as the Rainbow Snake, continue to inhabit the landscape, and may be reincarnated as its human inhabitants, tying the ancestral spirits, humans and the land into one system of belief.

During the last millennium, the Polynesian peoples settled on several thousand islands in the Pacific Ocean, in a vast triangular area whose points are New Zealand in the south, Easter Island in the east and Hawaii in the west. The North and South islands of New Zealand were the last places to be settled on earth, being reached around 600 years ago. Despite this enormous geographical dispersal, Polynesian beliefs retained a remarkably uniformity. Most Polynesian cultures believed that the gods are the children of the union of the male sky (known as Rangi by the Maori of New Zealand) and the female earth (known as Papa by the Maori of New Zealand). The gods each have their own specific domains and functions, like the deities of Greece and Rome. The leading god of the Maori pantheon is the god of the forest, Tane, who is believed to be the ancestor of the human race.

Papa and Rangi Parents of the gods

Rangi and Papa were the divine parents of the gods of the Maori of New Zealand.

Rangi and Papa were the primeval beings who created the Maori's six major gods. In the beginning was the absolute void out of which emerged two beings, Rangi, the sky, and Papa, the earth, who were locked in a tight embrace. They produced six children: the gods of the forest, the sea, war, cultivated plants, wild plants and the elements. But so tight was their embrace that the newly-born gods were trapped between their parents, unable to move or breathe. One of the gods cut off Rangi's arm, which allowed his brother, the god of the forest Tane to use his body to prize his parents apart. Once this had been done, the gods used wooden props to hold the celestial couple apart. The rain is said to be Rangi's tears of grief at his separation from his beloved Papa. Some of the gods chose to remain with their father in the heavens, while others, stayed with their mother, the earth.

DREAMTIME

Mythical period of creation when ancestral spirits of great power emerged from the ground, descended from the sky or rose out of the sea. They gave the earth its shape and natural features, often using their own body parts to do so. They created the animal species, human cultures and gave everything a name. These beings are themselves called "dreamings," and they provide the Aboriginal peoples of Australia with their personal identities through direct descent and a link to the land that they created.

Rainbow Snake Spirit of the waters

One of the most common ancestral spirits from the "Dreamtime," the Rainbow Snake embodies the male principle in Aboriginal beliefs.

The most common of the ancestral spirits found in Australia, the Rainbow Snake is a symbol of water and its replenishment during the rainy season. Because Australia is mostly desert, the search for water is a never-ending quest for the Aboriginal peoples and central to their myths. The snake's coils are reminiscent of the shape of a river across the landscape, and also of the rainbow. Like the waters, the Rainbow Snake is life-giving but it can also be destructive and dangerous to humans. In many stories, the Rainbow Snake devours humans whole. A phallic masculine symbol, it is often paired in myths by the feminine principle, such as that embodied by the Wawilak Sisters.

Tane Lord of the forests

One of the six sons of Rangi and Papa, Tane is the Maori god of the forests. He created light in the form of the sun and moon, and made the stars to clothe his father's body. He is hailed as the first planter of trees. His first attempts were not entirely successful. He made the branches legs for the trees to walk on, and the roots like hair, which waved in the open air. At length, he turned the trees upside them and they took root. Henceforth they would provide food for humans and animals.

Tane argued and fought with his jealous siblings and was promiscuous, mating with many female beings to produce animals, natural features and plants. On the advice of his father, Rangi, he made the first woman, with whom he at once mated. She bore him a daughter with who he also had sexual intercourse. The girl was so horrified by this incestuous liaison that she fled to the underworld, vowing to take the children of Tane with her, thus creating death.

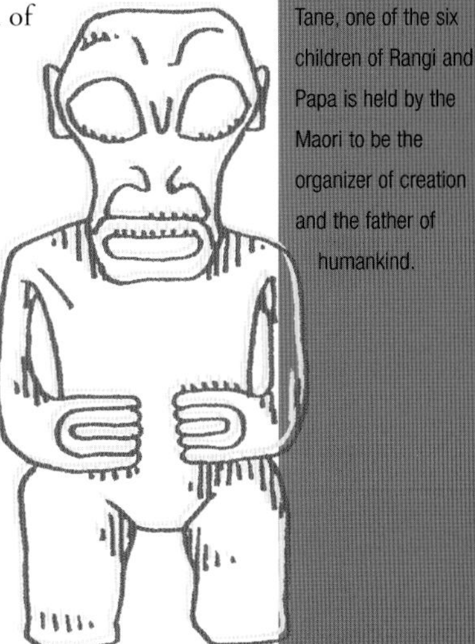

Tane, one of the six children of Rangi and Papa is held by the Maori to be the organizer of creation and the father of humankind.

Wawilak Sisters Female ancestral spirits

The Wawilak or Wawalag Sisters are examples of the feminine principle that balances the masculine principle of the Rainbow Snake. During the Dreamtime, the sisters emerged from the sea off Arnhem Land. The elder carried a small child, while the younger was pregnant. They both carried spears, which they used to hunt the animals they encountered with great skill. They made their way across the land, naming the animals and plants they encountered, bringing meaning and form to what had formerly been formless. When it was time for the younger sister to give birth, they stopped by a waterhole. The elder sister went hunting but any animal she tried to cook, leaped out of the fire into the waterhole. Unbeknown to the sisters, the waterhole was the home of the Rainbow Snake, who owned all the game animals in the area. Disturbed by the sisters, the snake swallowed them whole. However, he was later forced to vomit them up. The story deals with the origin of the rainy season, which is the product of an interaction between the snake, the masculine principle, and the sisters, the feminine principle, who alone are not able to bring the rains.

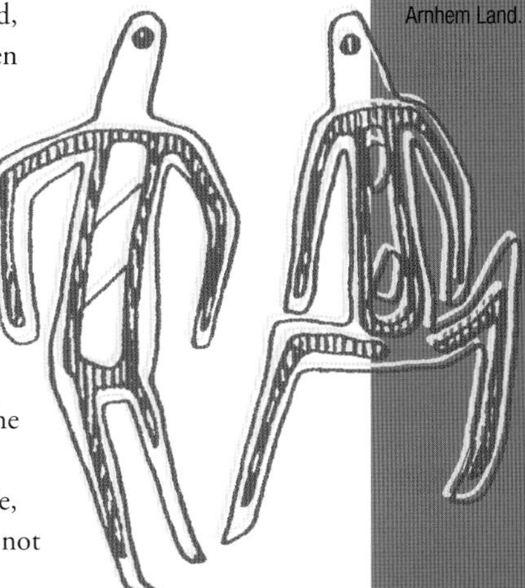

A symbol of the feminine principle, the Wawilak Sisters are ancestral spirits of the Aboriginal peoples of Arnhem Land.